No Greater Faith than that of Science

Manuel Vergara

Copyright © 2020 Manuel Vergara
All rights reserved. No part of this publication may be reproduced, distributed, or transmitted in any form or by any means, including photocopying, recording, or other electronic or mechanical methods, without the prior written permission of the publisher, except in the case of brief quotations embodied in reviews and certain other non-commercial uses permitted by copyright law.

The Library of Congress has cataloged the hardcover edition as follows:

Manuel Vergara
No Greater Faith than that of Science. – First Edition.

ISBN 978-1-7352891-0-6 (trade paperback)
ISBN 978-1-7352891-1-3 (Ebook)

Published in the United states by Manuel Vergara

Cover designer and Illustrators: Fajar Wahyu S. B. Bravoboy and Manuel Vergara

Editors: Cristine Taylor, Aura Vergara and Carlos Bravo

Https://www.nogreaterfaith.com

*For those who never stop asking the deep questions of life, and on looking for answers embrace Science ... and for those whose inner nature made them wonder beyond matter to
transcendence…
and for those who hold strongly to both!!!*

Table of Contents

Introduction	3
Evolution	21
Consciousness and the Self	43
Difference between Religion and Spirituality	83
Final Considerations	93
Appendix	101
Endnotes	103

Introduction

It's amazing the feeling of being awakened by music...when you hear a masterpiece like Chopin's Concerto No. 1 in E minor, or Rachmaninoff's Concerto No. 2 in C Minor or a lesser known piece as the Intermezzo No. 2 "Lejano azul- Far Blue-" from the Colombian composer Luis A. Calvo, it is impossible not to feel the emotion, the wonder, the chill, the peace, the tranquility, and the lifting of the mind; it seems that when the sound of some chords physically and literally touch you inside your heart, they awaken in you the excitement of crying or joy; other sounds like the third movement of Chopin's same concerto cannot be described as anything other than "heavenly." This makes us think with our inquisitive mind about how it is possible that a physical phenomenon such as the vibrations of matter (hit the piano strings) that travel through the air and create resonant effect in our tympanic membranes, then transform from kinetic to bioelectric energy, then are transmitted to our upper temporal auditory cortex[i] (see Fig. 1), and then shared and processed by many other parts of the brain can generate a deep and spiritual feeling, which is by the way very different and subjective in each one of us...it's very possible that the reader does not share my taste for music and does not like the pieces above; however, I am sure that you have your own that make you transcend.

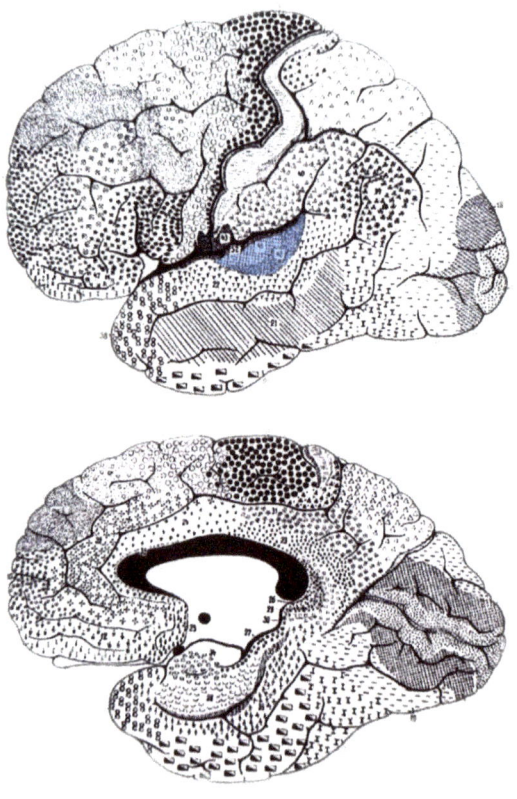

Figure 1. The modified (final) version of the Korbinian Brodmann's area map. Published in 1910. Brodmann was a German Neurologist who classified the cerebral cortex into 52 areas based on its cytological architecture. Auditory cortex areas 41 and 42 are highlighted here in blue. From Brodmann K., Zweiter Abschnitt: Physiologie des Gehirns, In: Von Bruns P. (ed.) Neue Deutsche Chirurgie, 11. Band: Allgemeine Chirurgie der Gehirnkrankheiten. I. Teil, Stuttgart, Verlag von Ferdinand Enke, 1914, 87-426.

The transformation from something physical to a transcendental experience seems wonderful and it is almost magical, and it is difficult to explain how something purely physical and material can cause such a lifting of the spirit. We know the anatomy and physiology, the location of the receptors, the brain connections involved in sound transmissions, but we do not know why that physical sound creates a profound unique, individual, and transcendental experience.

The same can be said of transcendence, feelings and emotions when seeing a sunrise, a sunset, and being able to appreciate those colors and contrasts, the slow changing movement of the clouds, the different shades of color and how they are modified with the position of the sun. Of course, there are physical and atmospheric dynamic effects of light change... but the question that arises is: why is it that what we see, we perceive it as beauty? It makes you wonder....

What could we say of a painting, although there are infinite combinations of tonalities, shadows, lights, and perspectives, it is a language that can convey feelings of exaltation, admiration, sadness, happiness, rejoicing, and sorrow, and send you completely to the abstract or make you feel like you are part of the scene in the painting. It's certainly more than contrasting tones, light, and shades... there's a feeling of emotion!!!

Spoken and written language is also incredible. How can your heart be filled with deep sadness and then rejoice in reading the Colombian poet Jose Asuncion Silva in his work "Una Sola Sombra- A single shadow-" Nocturno No. III (referenced at the end of the chapter). The spoken word can hurt you deeply and leave indelible marks on your heart, and the word can comfort you, lift you, create revolutions, destroy empires and change the lives of many generations to come as in the case of scientific leaders who have had and have as much influence on how we see life, the world, and the universe. It is the language that in great part makes us humans and makes us able to transcend when we share our expression with the rest of the world.

Since my school years, I have always had a curious mind, and I have always wondered why things occur, and I wanted to know how things work. I have always marveled at the incredible tenacity of thinkers and scientists of our history since most of them without any of the technological advances that we have today saw the same observations as we see them presently. At school I could see Jupiter's satellites with a homemade PVC telescope as professor Galileo Galilei did, but it did not cross my mind that I could calculate the speed of light as did the Danish astronomer Olaus Roemer (Fig. 2).

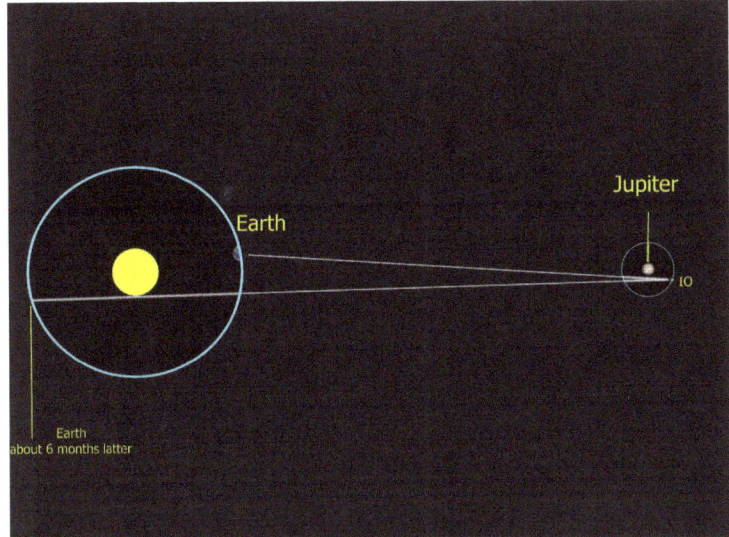

Figure 2. In About 1675 Danish astronomer Olaus Roemer was measuring the period of IO which gets eclipsed by Jupiter as seen from Earth once every orbit; by carefully measuring these eclipses he noted irregularities with the interval between these successive eclipses which were longer and then shorter when Earth in its orbit was moving away and towards Jupiter by about 22 minutes. He reasoned that these must be the time light takes to travel the diameter of Earth's orbit. By estimating the Earth's orbital diameter, he gave the basis for calculating the speed of light as 220.000Km/second (not bad considering it was the 17th century) which was done at that time by Christiaan Huygens.

We see objects fall all the time, but it does not occur to us that we could determine the law of gravity that governs these objects and the entire cosmos, we see the colors of the rainbow, but we do not initially conceive of them as the decomposition of white light as Newton did (Fig. 3).

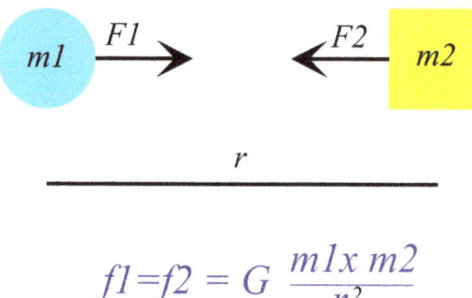

$$f1 = f2 = G\,\frac{m1 \times m2}{r^2}$$

Figure 3. Newton's main contribution: the law of universal Gravitation depicted above with G directly measured 71 years after Newton's death in 1798 by Henry Cavendish and obtained a G= 6.674 x 10-11 m3.kg-1.s-2. Newton contributed significantly to the field of light and optics including white light decomposition into its visual spectrum and also, He built a first practical reflecting telescope which still carries his name .

We all see varieties of birds and breeds of dogs, and yet it does not cross our mind to understand the theory of evolution as Darwin did (Fig. 4).

Figure 4. Finches from Galapagos Archipelago. From Darwin's Beagle Journals 1839.

Through their brilliant minds, those scientists could see beyond the apparent world and make incredible discoveries that were then embodied in amazing observations and descriptions.

There are other qualities in nature that speak of transcendence far beyond the material, the physical, and the physiological such as a mother's love for her son or daughter, although it is true that it can be modulated as it can be much of our behavior by prolactin and vasopressin (Fig. 5), cultural and social learning, the presence of mirror neurons, genetics, it impact us much more than a biological entity predetermined by hormones, neurotransmitters, and learning; this love seems to transcend the physiological, and it is a type of love that can sacrifice itself which goes against the most powerful mechanisms of self-protection and survival common to all species. That kind of love speaks of something beyond the physiological being.

Figure 5. Prairie Vole (Microtus Ochrogaster), classic example of only about 3% of mammals which are monogamous, and permanent bonding occurs after the first mating. Male voles also guard the nest and help to rear the pups. This is in contrast with the mountain vole which is promiscuous and does not take care of the pups. It was found that the main difference is that the prairie vole has a higher density of receptors for Arginine Vasopressin and Oxytocin in the Ventral Pallidum and nucleus Accumbens (brain nucleus related with motor control and rewarding) than the mountain vole. (C.S.Carter et al. "Oxytocin, Vasopressin and Sociality".Progress in Brain Research.170,2008,331-36).
There are also subtle differences in the Vasopressin receptor gene. Although this aspect is not the entire picture that determines monogamy and bonding, it is a major neurophysiological factor. From Creative Commons by phenology.

It is more acceptable to think of a mother who gives her life for her child, but it is even bigger – albeit rare – that someone sacrifices her/his life for anyone other than her family member. This is not so uncommon as it occurs in other species such as ants, vervet monkeys (Fig.6),

Figure 6. Vervet monkeys show altruistic behavior by giving alarm calls to warn fellow monkeys of the presence of predators, even though in doing so they attract attention to themselves, increasing their personal chance of being attacked. Image from Creative Commons, by richardrichard CC BY-ND 2.0 License

and meerkats to mention a few, however, this behavior in these animals is fixed and determined at the genetic level. In humans, behavior is a choice and certainly goes beyond genetic self-determinism. That choice is what we call free will, which makes us human and again speaks to us of some goodness hidden deep within us that speaks of our connection with our origin.

I have always been interested in the Buddhist community and its basic teachings. The Buddhists are very detached from material things (attachment and desire are the cause of suffering) and therefore without them, they have a very full spiritual life. I am surprised that there are Tibetan Buddhist monks who have dedicated their lives to meditation and contemplation and even Tibetan teachers like Mingyur Rinpoche who have created an entire school of meditation and are spreading it through western countries as an effort in the search for a spiritual life through meditation can provide happiness without the need for material things on which we depend so much in the west.

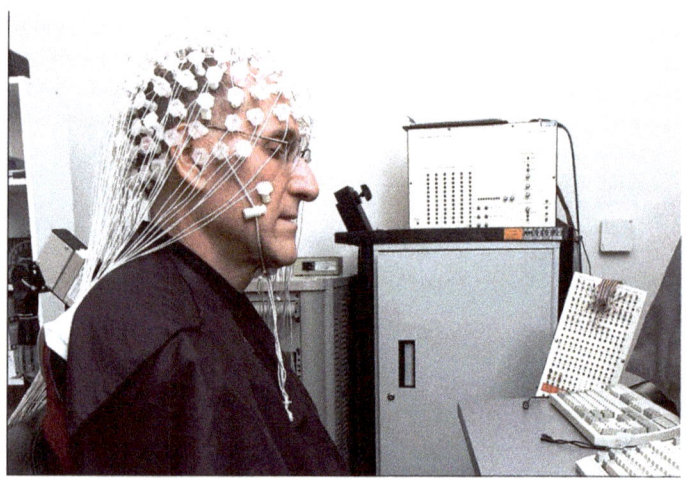

Figure 7. Expert meditator Barry Kerzin having an electroencephalogram recording during meditation for neuroscience research. Dr. Kerzin is an American physician and affiliate professor at University of Washington Tacoma as well as an officially ordained Buddhist monk by the

Dalai Lama in 2003. He is the founder of Altruism in Medicine Institute with the vision of emphasizing training in meditation, mindfulness, and compassion for medical doctors, nurses, and related health professionals. Picture from Wikimedia Commons, Public Domain.

By the way Electroencephalogram (EEG) and Functional MRI (fMRI) studies have demonstrated that in master meditators such as Dr. Kerzin (Fig. 7) the Gamma oscillations (greater than 30 Hz frequency of the brain waves) are a far more prominent feature of their brain activity than in other people. The contrast between the expert meditators and controls in the intensity of gamma was on average 25 times greater amplitude gamma oscillations during baseline compared with the control group. It is interesting and also mind-boggling that the gamma activity remains in the meditator even during sleep!

Using fMRI, it was identified that key reductions in the posterior cingulate hub of the DMN -Default Mode Network-, (Fig. 8), along with increases in right frontal and left temporal areas, in experienced meditators during rest and during meditation in comparison to healthy controls (HCs), and the duration and frequency of occurrence of DMN microstate was higher in meditators compared to HCs.

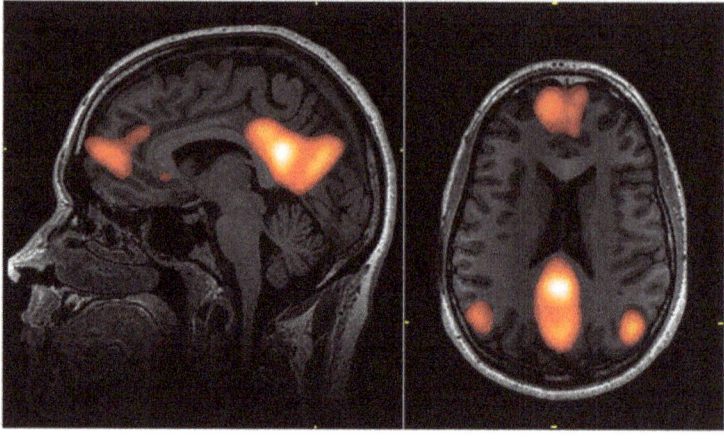

Figure 8. Default Mode Network (DMN). This is a spontaneous brain activity network which is active during passive rest as when we are thinking about self, about others, remembering the past, and envisioning the future, and usually gets deactivated during attention-demanding tasks such as visual attention or cognitive tasks. The DMN includes the Precuneus, anterior and posterior Cingulate cortex and bilateral parietal inferior gyri. During meditation as noted there was a significant deactivation of the posterior Cingulate region with increased activation on right frontal and left temporal regions.
From John Graner, Neuroimaging Department, Walter Reed National Military Medical Center, Bethesda, MD, USA. Public Domain.

The interactions between more evolved animals such as mammals and primates make us think about what exists outside the evolutionary "box" and the struggle for survival. We have all seen how in these species, there is room for play as with families of dolphins, or the play of cats and felines, or space for relaxation in community such as chimpanzees delousing each other or even closer to us, the happiness expressed by our close companions dogs when playing with their owner or when seeing them arrive home. All this tells us that there are existential motives and goals beyond the raw evolutionary survival ... there is happiness and joy in living.

There are many other human qualities that elevate our condition that seem only material at first sight. We have imagination, premonition, good clinical sense, empathy, creativity, self-questioning, kindness, affection, curiosity, and the search for answers and knowledge.

All of the above shouts out something tangible that is our spiritual nature, which is much beyond and bigger than our physical body and the physical phenomena, not only in our species, but in many others, especially the most advanced in the evolutionary process, and to a lesser degree would extend to all other species. It is very likely that through the evolutionary process we have acquired consciousness which in turn allows us to realize our spirituality.

This matter of consciousness is the great scientific mystery of our century. It implies the fact that we question ourselves... we are aware of being conscious (conscious awareness)! Is it perhaps that consciousness is an illusion or a consequence caused by neural networks?, are we only robots without free will? Or is it more than that?

In addition, consciousness and the fact that you are aware of being aware, whatever its nature, cannot translate into physical connections because it is a unique and a subjective experience for each person. Consciousness allows a deeper concept: the concept of the Self that makes us unique in the entire universe. Later a chapter to this important topic of Consciousness will be dedicated.

Nocturno III, Jose Asuncion Silva

One night

A night all full of perfumes, murmurs, and music of wings,

A night

on that burned in the bridal and wet shadow, the fantastic fireflies

next to me, slowly, against me all tight,
mute and pale as if a feeling of infinite bitterness,

to the most secret bottom of your fibers would shake you,
by the path that crosses the blossomed plain

you walked

And the full moon
through the blue, infinite, deep skies spread its white light,

and your shadow

thin and weak,

and my shadow

by the rays of the moon cast

upon the sad sands

of the path, gathered together.

And they were one

and they were one

and they were one long shadow,

and they were one long shadow,

and they were one long shadow!

This night

alone, the soul

full of infinite bitterness and agony of your death,

separated from yourself, by the shadow, by time and distance,

by the infinite black, where our voice does not reach,

alone and mute

by the path I walked,

and one could hear the barks of the dogs to the moon

to the pale moon,

the squeal of frogs,

felt cold,

it was the cold they had on your bedroom

your cheeks and your temples and your adored hands,

among the snowy whiteness

of the mortuary sheets!

It was the cold of the tomb, it was the cold of death,

It was the cold of nothingness...

And my shadow

by the rays of the moon projected,

was going alone,

was going alone

was going alone through the lonely steppe!

And your slender and agile shadow

Thin and weak

as in that warm night of the dead spring,

as on that night full of perfumes, murmurs, and music of wings,

came up and marched with her,

came up and marched with her,

came up and marched with her!......

Oh, the bound shadows!

Oh, the shadows of the bodies that join together with the shadows of the souls!

Oh, the shadows who seek each other and join in the nights of sadness and tears!

Evolution

The variety and richness of life on earth is incredible. How did there get to be so much diversity of life forms on Earth?

It is really hard to appreciate evolution in our daily lives, we don't really see it. Yes, we agree we know about the large variety of dogs and cats and of plants and fruits and despite the great difference between a Great Dane and a Chihuahua, we realize that they are still of the same species, they are still dogs. We learn in microbiology that it is not uncommon for bacteria to have spontaneous mutations that make them more resistant to antibiotics or that they can use different sources of energy (Fig. 9.), we still know that they are still bacteria, and we don't see them becoming a different species. In the course of a human life, we cannot observe macroevolution, which is the change from one species to another, since it takes thousands if not millions of years to occur.

Figure 9. Richard Lenski's long-term evolution experiment with E. coli taken on June 25, 2008. Professor Lenski had taken in 1988 twelve

asexual identical populations of E Coli and has followed the genetic changes which occurred over a total of 66,000 generations by November 2016. He has found different phenotypic and genotypic changes including faster growth rates and increased cell size (volume increased by about 2 to 2.5 times) and also increases in the rate of mutations. The most striking finding was the evolved ability of aerobic growth on citrate – see above A-3 population — is more turbid from been able to grow in citrate medium-(normally grows in glucose), which is unusual in E. coli, in one population at some point between generations 31,000 and 31,500. From Wikimedia Commons, Attribution-Share Alike 1.0 Generic license

The reason why our direct experience in life does not allow us to observe the change from one species to another is because our life on earth is very short and ephemeral, like a ray of light that ignites in the darkness and before we know it is extinguished. This can be compared to the age of the sun, the Earth and of the universe. It is surprising that even with a minuscule life, we can question the nature of time.

First, we have to realize that the current estimate of the age of Earth is 4.5 +/- 0.05 billion years (4,500,000,000 years). This has been corroborated by radiometric estimates analyzing rocks, as well as analyzing meteorite, moon, and Mars rocks as well. Radiometry is based on the fact that some radioactive elements in the rocks become other elements with the passage of time at a predictable speed given by their half-life; thus, uranium is transformed into lead, potassium into argon and strontium into rubidium. By estimating the proportions of these pairs of elements in any rock, you can estimate its age (Fig.10).

Figure 10. This are natural Zircon gemstones and Zircon Crystals ($ZrSiO4$) under microscope. Zircons are special crystals which are found in many rocks and contain traces of other elements including Hafnium, Phosphorus, Yttrium, Thorium, and Uranium, but has no Lead at all; so all the lead over time found on Zircon is caused by decay of Uranium, so by calculating the proportion of U and Pb we can figure out how much time has elapsed. About 1mg of this is obtained after crushing a 10 Kg rock, then pulverizing it, then separating the crystals by mechanical and fluid density features and then by passing them through a magnetic field that separates iron containing minerals from

zircon and then by selecting best individual crystals by hand under a microscope and then dissolving them with Hydrofluoric acid into a solution which then is filtered further to eliminate all the other elements from the solution other than the Uranium and lead (at this point Lead measured in Picograms:10^{-12} g).Then the proportions of Uranium and lead in the sample are measure on a Mass Spectrometer in which the sample is heated to about 1500 Celsius and evaporated and also Ionized (lose an electron) and in that way becomes positively charged so they could be focused and then accelerated by a high voltage and then exposed to a high electromagnetic field and deflected angularly according to their respective masses which allows identification and quantification of specific elements by a detector. Then you may calculate the age of the rock. The half-life of an element as Uranium is determined by measuring the number of Alpha particles (helium nucleus) that are released per second and then derived from standard formula. From Shutterstock.com and Stock.adobe. com.

When one reads the Origin of Species, one can only realize the greatness and quality of the biologist. He is seen as a very observant being with great attention to detail, having the ability to make incredible descriptions of his observations; a very special mind, similar to those of Copernicus, Newton, Galileo, and Einstein who were capable of discerning very profound mechanisms about the laws of nature.

Darwin also failed to observe macro evolution during his several years voyage on the Beagle boat. Darwin was able to see variations in the finches but they were still finches, etc. However, he was able to consider the fact that given sufficient time, one species could change and transform into another. His observations in embryology, fossils, and competition for existence supported his theory of Natural Selection by small series of variations and modifications that would originate different species. He went even beyond this premise and stated his hypothesis that all species that live or have lived could originate from a few or a single common ancestor.

There are several scientific facts that validate the Theory of Evolution, to mention a few:

The fossil record: fossils show a progression of simple shapes which are older than more complex forms that are more recent. The earliest fossil is in Australia—stromatolites that were able to survive 3.5 billion years because the bacteria that produced them produced calcium carbonate and turned into rock. Otherwise it could have never survived a vestige of unicellular life over such a long time.

After this, it has been documented that in the so-called Cambrian Explosion 500 million years ago, the best preserved fossils include those found in the Burgess in the Walcott Quarry in British Columbia (Fig. 11) where special conditions were given for the preservation of fossils, when a large mountain of mud detached and buried thousands of marine creatures that were trapped below.

Figure 11. Cambrian Trilobite Olenoides Mt. Stephen (British Columbia). They lived in the Cambrian period about 500 million years ago and became extinct about 250 million years ago. By Wilson 44691, Wikimedia Commons. Public Domain.

The fossils (Fig. 12) also show the similarities that exists in all vertebrates: many mammals have seven cervical vertebrae no matter if they have the neck
as long as a giraffe; the wings of a bird, the wings of a bat

and the arm in a human they are all similar; also many mammals including humans have vestigial organs that do not seem to have a necessary function like the appendix; also the trajectory of the Recurrent Laryngeal nerve, branch of the Vagus Nerve (which makes a rather long one-way route and then returns to finally reach the Larynx) is maintained among many vertebrates from fish to mammals.

Figure 12. Platypterygius Sachicarum. Is an ichthyosaur of the family Ophthalmosauridae. Lived in the early cretaceous era (about 145 to 100 million years ago) in a cosmopolitan distribution including the Paja formation in Central Colombia. With permission from Petter David Lowy Ceron Director of the Museo Paleontologico de Villa de Leyva, Colombia. Sciences faculty of the Universidad Nacional.

There are intermediate forms such as the Archaeopteryx (with characteristics of bird and reptile),the Tiktaalik a fish with transition shapes between fins and legs that could walk on the ground, the Ambulocetus that was a whale but with feet, toes and fins, and to go closer when looking at the life phases of the salamander or the regular frog, you find characteristics of water fish in the larva phase (with gills and long tail for swimming as tadpoles) and then there is a metamorphosis that converts them into a tetrapod with legs to walk on land as an adult.

Comparative embryology (Fig. 13) identifies the similarities between embryos of different species such as mammals, birds, lizards, and snakes which are extremely similar in their earliest states and it is difficult to differentiate them, suggesting that they originate from a common ancestor and that their embryonic development began equally, and only in more advanced stages of development they differ more and more until they are completely different at birth.

Studying human embryology in the faculty of medicine, I was always puzzled by the fact that in the stages of early human embryonic development (weeks 4-6), we look like fish embryos and even develop branchial arches (Pharyngeal arches); which later in fish originate the branchia and in humans originate structures related to the mandibula, hyoid bone, pharynx and larynx. Another example is how we develop a tail at the end of the column in early stages, which then reabsorbs and disappears later in embryonic development. Darwin, analyzing these similarities in embryonic development, concluded: "Community of embryonic structure reveals community of descent" (*The Origin of Species*, 1859). These changes that occur during embryonic development are regulated by genes such as the Hox genes which determine the embryonic development of the limbs in many different species including humans.

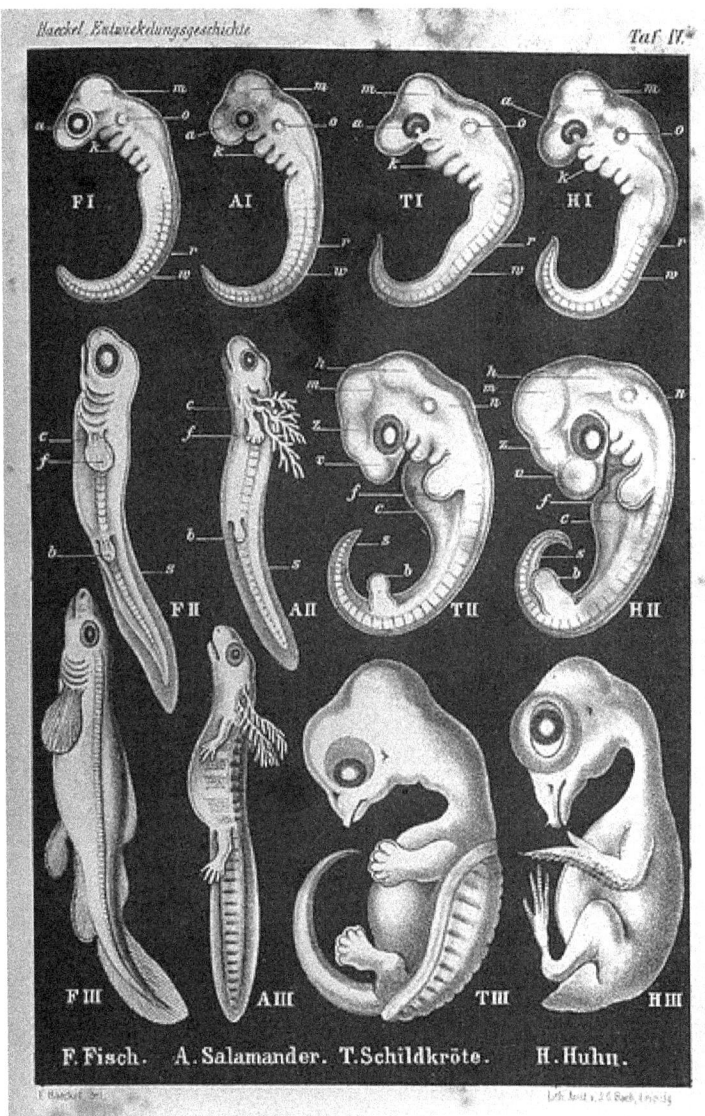

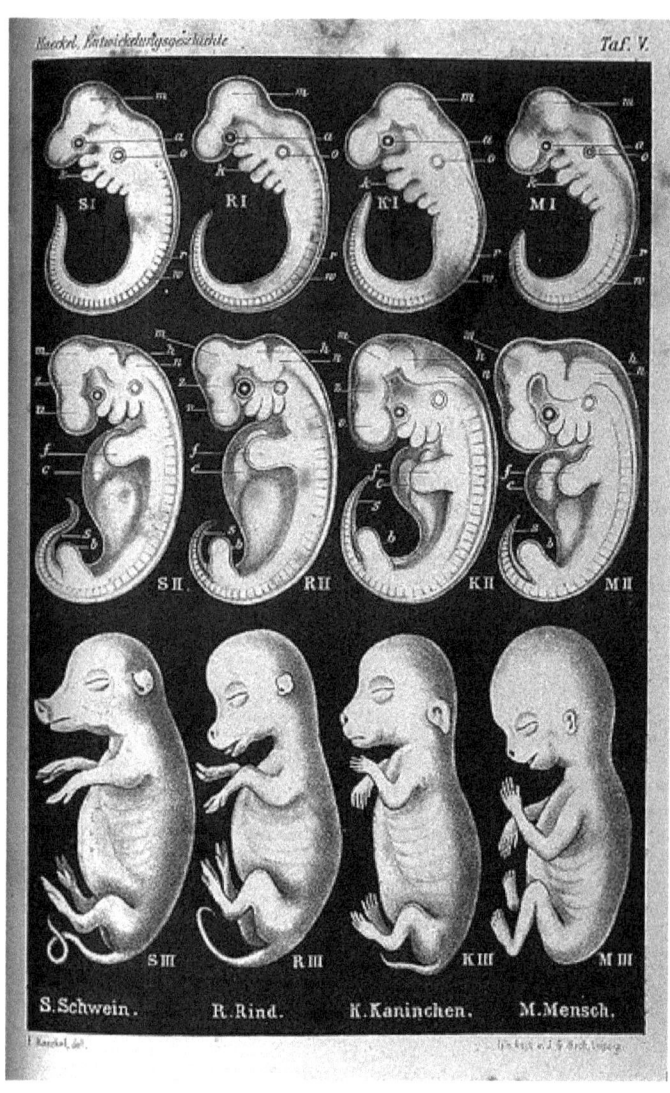

Figure 13. Original drawings from German Biologist Ernst Haeckel of comparative embryology of multiple species including fish, salamander, turtle, chicken, pig, cow, rabbit, and human. As noted, the initial stages occur at about 4 weeks in human development and show striking similarities with the presence of pharyngeal arches and a tail.

He proposed that the Ontogeny (individual embryonic development) recapitulates the Phylogeny (evolutionary development of multiple species). Haeckel was not without controversy since he was accused of fraud with some of his drawings and also, he believed that although humans had a common ancestor later the current different races evolved separately, which later had political influence on the Nazi philosophy. By Ernst Heinrich Haeckel, Creative Commons attribution 4.0 License.

Perhaps the greatest evidence in favor of evolution is given by genetics. As we know there has been an exponential development of this science in the last 30 years. One of the greatest achievements has been to be able to obtain the sequence of the human genome completed in April 2003. It brought many surprises to the scientific community who know now that the human genome is made up of 3.200 million base pairs or codes (6.4 billion in the body's somatic diploid cells) and only 1.5% of these (20000 to 25000 genes) are encoded for protein synthesis. More recently it has been documented that some genes do not encode proteins but express regulatory RNAs which has brought the list to about 47000 genes.[iv] Of course, the genome of other species was and is being sequenced, and of course we like to compare it to that of ourselves. Starting with ourselves, it is known that no matter the great differences between the human races, all humans are 99.9% identical. By comparison we are 98.9% identical to chimpanzees. The number of genes and chromosomes was thought to be higher in more advanced species like us, but surprisingly this is not the case: we have 46 chromosomes and the rhesus monkey has 32, the chimpanzee, the gorilla, and the orangutan have 48, ants 2, zebra fish 50, dog and wolf are 78, a lichen of the group Ophioglossum has 1262! A japanese flower, Paris Japonica may possess the largest known genome of any living organism, although having only 40 chromosomes, it has 150 billion DNA base pairs.

If we take the human genome and lengthen it and place it in a horizontal line and then compare it to the genome of other species, we will find very significant similarities much more in the genes that encode it by a protein (exons) than those that do not (introns). When making this comparison, we see that we are 100% similar to the chimpanzee in terms of exons, and 99% similar to a dog or a mouse, but only 75% similar to a chicken and 60% to a fruit fly and 35% to an earthworm. The similarities almost completely disappear when we compare the human genome with species other than mammals in terms of introns or the sequences between the genes.

If we do this with many species and compare their genome, it is possible to build a "tree of life," (Fig. 14) which points to a common ancestor or a few common ancestors from where the original genetic base originated and then had progressive variations originating other species. Life organisms are classified taxonomically in Domains which include Viruses, Archaea, Bacteria, and Eukaryotes-cells that have nuclear membranes-. These in turn are classified into Kingdom, Phylum, Class, Order, Family, Genus, and Specie. This tree of life based on genetic comparisons is very similar to the tree of life based on fossils and by comparative anatomy and embryology. The variation of genes occurs due to mutations occurring in the order of arrangement of the letters of the genome (base pairs). Mutations are estimated to occur approximately once every 100 million bases of base pairs; although it would seem uncommon, due to the large number of base pairs in humans (3.200 million from each parent), we have approximately 64 new mutations for each individual in their lifetime.[v] This is the mechanism and process that operates in the "Natural Selection" postulated by Darwin which he did not know during the time when he published the *Origin of Species* in 1859.

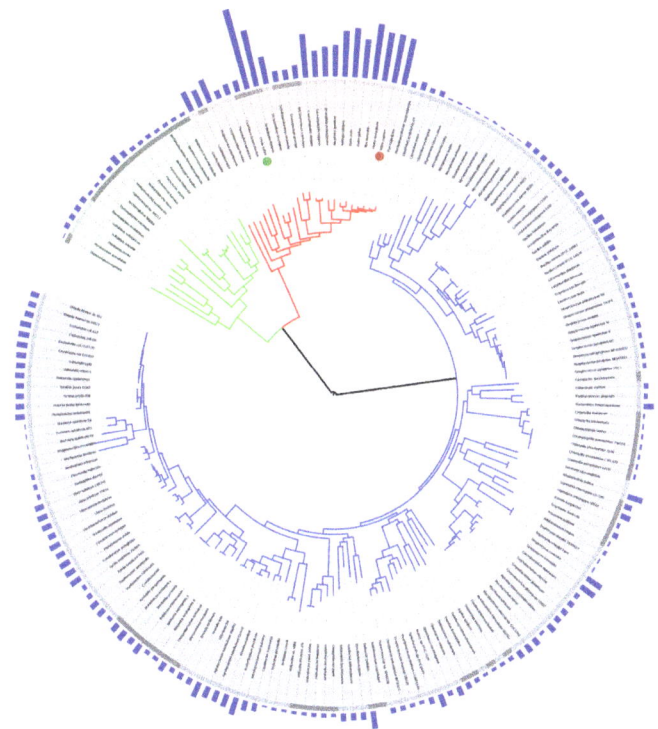

Figure 14. This is a genetic tree of life based on the alignment of 31 universal protein families and covers 191 species whose genomes have been fully sequenced. Three main groups are defined: Bacteria in blue, Archaea in green, and Eukaryotes (cells with clearly defined Nucleus) in red. Make a note of Homo sapiens (red circle) and Oryza Sativa -

rice- (green circle) as examples within the Eukaryotes of animals and plants. The outer bars represent the genome size. From Ciccarelli, FD: "Toward automatic reconstruction of a highly resolved tree of life", Science 311(5765): 1283-7, 2006. Public Domain.

There are some genes that are common to humans and chimpanzees like Caspase 12 (the gene is located in the same position in the genome of both species) and a thorough mutation became non-functional in humans. The protein-coded gene with inflammatory functions in humans of non-African descent was inactivated which apparently offers greater resistance to sepsis.[vi] Another gene the Myh16 encodes by a muscle protein Myosin Heavy Chain 16. This gene is functional in primates which explains its powerful chewing muscles; in humans this gene mutated and became non-functional which is postulated as an explanation for which masticatory muscles (temporalis and masseter mainly) would allow the growth of the head and brain in humans.[vii]

These examples of genes that are in both genomes of humans and primates which had a defined function but that in humans mutated, converted, and became non-functional genes suggest that these changes have occurred by natural law and not by special and independent creations by God. The same is suggested by the fact that certain anatomical changes seem to have no clear explanation or may be better designed than they are; the classic example is the Inferior Laryngeal Nerve (branch of the Vagus nerve), which is embryologically originated in the sixth gill arch and which in vertebrates innervates the muscles of the larynx except for the cricothyroid. The Vagus nerve originates from the brain stem, goes down to the neck, and as it passes by the larynx originates the Superior Laryngeal Nerve that innervates only one muscle of the larynx: the Cricothyroid; the nerve continues to descend on the right side to the lower side of the subclavian artery and on the left side below the aorta and there at that level originates the branch of the Inferior Laryngeal Nerve that turns back (hence its recurrent name and goes up to the larynx where it innervates all the other muscles of the larynx except the cricothyroid. This same route is done in fish where the trajectory is very short and in all the mammal vertebrates including as examples the giraffe in which the nerve has to turn back 4.6 meters and in the case of a dinosaur would have to turn back about 28 meters! (Fig. 15).

The argument is made that if there is a designer, it is possible that the design would have more direct anatomical routes depending on each animal or species, rather than a pattern that seems to be maintained and gradually change among species. On the other hand, the nerve also gives sensory, secretor, and motor branches to the high segments of the trachea and esophagus, so it is possible that there are other functions that justify their recurrent journey.

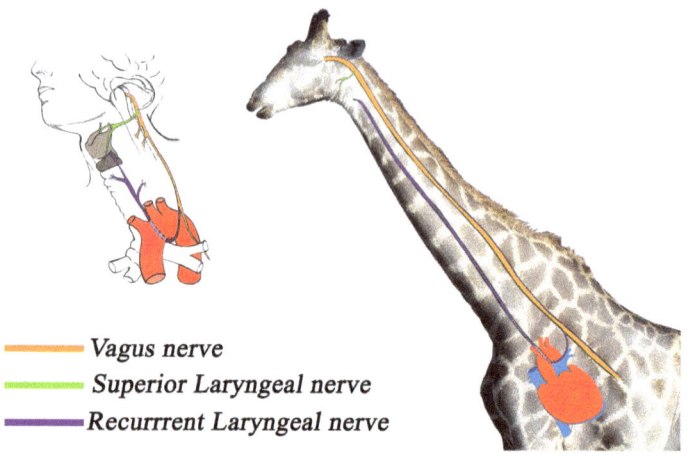

- Vagus nerve
- Superior Laryngeal nerve
- Recurrrent Laryngeal nerve

Figure 15. The larynx is innervated by two nerves: the Superior and the Inferior Laryngeal nerves, the latter is also called recurrent because it directly innervates the laryngeal muscles following their path down to the large vessels that originate from the heart and then turns upwards until reaching the larynx for its nerve supply. The recurrent Laryngeal nerve originates from the sixth branchial arch and therefore loops around the large vessels (the subclavian artery on the right side and the arch of the aorta on the left side). This pattern occurs in vertebrates including fish, humans, giraffes and would also apply to dinosaurs and therefore their recurring path would be almost 5 and 30 meters respectively! Illustration on the Left By Andrew Meyerson under Creative Commons Attribution-Share alike 3.0 license. Illustration on the right by Bravoboy.

The data mentioned above is consistent with *the Theory of Evolution but at no time does it exclude the concept of the Creator who initiated the process.* It tells us indeed the Creator does not individually design each species or every individual of the species, but rather has allowed natural laws to undertake the evolutionary process. One thing is the evolutionary process and another thing is the origin of it. As Darwin said in the conclusion of his book The Origin of Species in 1859: "Authors of the highest eminence seem to be fully satisfied with the view that each species has been independently created. *To my mind it accords better with what we know of the laws impressed on matter by the Creator*, that the production and extinction of the past and present inhabitants of the world should have been due to secondary causes, like those determining the birth and death of the individual. When I view all beings not as special creations, but as the lineal descendants of some few beings which lived long before the first bed of the Silurian system was deposited, they seem to me to become ennobled."

Regardless of the religious beliefs of each individual and regardless of their belief in God, it is important to consider the concept of the Creator's existence. *In the Jewish and Christian ideology, evolution is not in conflict with the Torah or the Bible* since Genesis does not say that God created each being individually but says, "produce the earth vegetation" (Genesis 1:11, "let the waters teem with living creatures" (Genesis 1:20, "produce the earth living beings according to their kinds" (Genesis 1:24, and "God formed the man from the dust of the ground" (Genesis 2:7). Very similar to the above mentioned by Darwin: "laws imparted to matter by the Creator."

The concept of a Creator is central to other eastern Ideologies: the Advaita Vedanta school of Hinduism which holds that there is only one causal entity (Brahman) and the Sikhism which believes in an Universal God who is the universal creator, sustainer and destroyer. Although Einstein did not believe in a personal God, He did believe in a God who revealed himself in the harmony of all that exists, similar to the God of Benedict Espinoza, philosopher of the 17th century who saw God and nature as one and as the same substance of reality.

Although the evidence is consistent with evolution as the source of the great biodiversity that has been and it is here on Earth, it is only reasonable to raise some unresolved gaps:

- If the occurrence of mutations is almost always to the detriment of the individual, why would not this change the balance towards destruction rather than towards evolution?

- Viruses are obligated parasites. If evolution has a direction from the simplest to the most complex, how would one explain the origin of viruses that could not exist or replicate in the absence of more complex organisms such as bacteria?

- In mammals that have few offspring, if from the genetic changes and mutations a new species is produced, how could that new species reproduce if it cannot be mixed with the previous species? This would be easily understandable in species with many hatches such as fish or insects because if another species is generated, there could be many individuals who could cross and maintain it, but this is more difficult for mammals with few descendants.

Many purely materialistic views claim that life originated randomly, by chance or out of nowhere, even that the entire universe originated from nothing which goes against any sense of rational thought (basic principle of no effect without cause); this has to do with the concept of the "nothing" that ultimately ends up not being nothing as we understand it semantically but it is what physicists call "quantum fluctuations" which are full of energy and fields and do not correspond to nothing.

It takes a lot of faith to believe in the origin of life by chance and by accident. Current theories of the origin of life moved from a "warm little pond" to deep sea floor hydrothermal vents where biological communities have been found living in this environment with no sun or light exposure but within reach of hydrogen, carbon dioxide, and sulphur coming from the geysers. Of course, humans the most intelligent species on the planet want to recreate life as well, and we have gone from the Miller-Urey experiment done at the University of Chicago in 1952 to recreate the conditions of the primitive Earth and used water, methane, ammonia, heat and electrical sparks which were able to produce organic compounds including hydrogen cyanide, formaldehyde ribose, and even amino acids, the building blocks of proteins.

Since then we have evolved at an incredible pace. The current Covid-19 Pandemic of 2020 has raised incredible interest in understanding viruses, and right after the pandemic started, there was full analysis of the Virus Genome -which consist of about 29.903 bases and 15 genes-[viii] by different countries trying to trace its origin and variability from country to country. This has been done for decades starting with the RNA Polio virus, which genome was generated with recombinant DNA technology in 1981.[ix]

At this time the full-length genome sequences of more than 9000 viruses are publicly available in a database maintained by the National Institute of Health. Unfortunately, despite all the damage they can cause, viruses are parasites and need a living cell to replicate. Efforts have been made to create a cell that can replicate on its own since this will amount to "creating life." In 2010 there was news of a team of researchers who had "created" artificial cells, "synthetic life," which is indeed a great accomplishment but was not the creation of life from scratch as they clarified later; what they had done was to synthesize and assemble the genome of a bacteria called Mycoplasma Mycoides and then transplant it into another recipient bacteria Mycoplasma Capricolum and create cells which are controlled by the synthetic chromosome and can replicate.[x] Later in 2016, they did the same experiment but reduced the number of genes that are indispensable for replication and sustainability to a minimum of 473 genes creating a minimal bacterial genome.[xi]

Of course, scientists are far away from really creating life from scratch even with all the knowledge and modern techniques in chemistry. Synthesizing the genome of viruses and bacteria is a big accomplishment, but engineering and recreating the machinery of a cell with all its organelles is another animal; yet there are efforts to recreate the parts of a cell from the bottom up, and there are teams already working on membrane cell-like structures or liposomes, or ATP energy producing mitochondria-like units. Anyone who has read a chapter on protein synthesis cellular processes will acknowledge the complexity of the task with many different cell organelles including the nucleus, the Golgi apparatus, the Endoplasmic Reticulum, and the Ribosome. Even if in the future it is possible to build a cell based only on synthetic chemistry, it will only demonstrate that you need a designer and a creator for this, and it will demonstrate that the assemblage of a living cell won't occur by chance. Some may say that if billions and billions of years are given everything is possible, nevertheless, the possibility for creating life by chance is as absurd as to have a million monkeys able to type continuously for billions of years and produce Shakespeare's *Romeo and Juliet* or Garcia Marquez's *One Hundred Years of Solitude*!

There is no Greater Faith, I would say, from those who can conceive that since the formation of the earth 4,500,000,000 years ago until the formation of the first single-celled organisms a billion years later and up to the formation of the first multicellular organisms about 600 million years ago (a total wait of 3,9 billion years) by a casual and miraculous combination of elements life will form on earth. It is very difficult to conceive that magnitude of time being our life as ephemeral as a shooting star. Science has always considered religion and the belief of a Creator as a matter of faith; however, theories about the origin of life as well as the origin of the universe are not falsifiable and are also a matter of faith. I would say as expressed by G.K. Chesterton (English writer and philosopher): "It is absurd for an Evolutionist to complain that it is unthinkable for an admittedly unthinkable God to do everything out of nothing and at the same time pretend that nothing turn itself into everything." On the one hand, chance and luck, and on the other hand a purpose imparted by the Creator. In matters of faith, the materialists are the winners here, there is no greater faith than theirs....

Consciousness and the Self

The materialistic and mechanistic philosophy of the brain holds that all our experiences including consciousness, the conception of the soul or spirit and any sense outside physical reality are pure fantasy originated in the intricate processes of the brain. This is based on several facts of which I will mention some:

Very familiar to neurologists is the fact that if there is damage in one part of the brain, then there is a change in function frequently that manifests itself as a neurological deficiency. It is very easy to understand that causing damage to motor parts of the frontal lobe cortex or its subcortical connections often causes weakness or paralysis on the face, arm, or leg of the opposite side (usually one side of the brain controls the opposite side of the body). In the same way, any damage to the Parietal sensory cortex causes decreased ability to feel sensations on the opposite side of the body.

Damage to the Fronto-Temporal region on the left side causes Aphasia, which is the difficulty to communicate and to understand spoken or written language. Language is a superior function of the brain that makes us human. If there is damage in the Occipital lobe, then there is loss or decrease of vision in the opposite visual field in both eyes (Homonymous Hemianopsia). More complex visual alterations can occur when, outside of having harmed the visual areas, there is damage to adjacent areas, including the inability to recognize familiar faces also called Prosopagnosia, which is caused by bilateral damage in the Temporo-occipital lobe or Fusiform girus ; Balint syndrome, which causes simultanagnosia (inability to recognize the entire landscape, only parts), oculomotor apraxia (difficulty fixing the eyes on an object), optic ataxia (inability to direct the hand to an object using a visual guide); Anton syndrome, also called denial of blindness, when there is bilateral damage of the Occipital lobes and adjacent areas. Damage to the middle part of the temporal lobes including the Amygdala causes hyperorality, hypersexuality, compulsive eating, visual agnosia (difficulty recognizing and interpreting visual information not due to deficit in vision), and docility.

Some patients have what is called disconnection syndrome, which is related to surgical procedures such as the resection of the Corpus Callosum (which connects the two brain hemispheres) and can cause apraxia of the left hand (the patient cannot perform complex movements - such as combing - with the left hand but has no difficulty with the right hand, and they cannot give the name of an object placed in the left hand but they can do it when it is placed in the right hand). Patients can also have the Alien hand syndrome (a hand that moves on its own without control, sometimes the right hand has to control the left hand and may have opposite movements), these patients retain however, the ability to make movements already learned with both sides, including swimming, cycling, or walking. Some patients with posterior lesions of the corpus callosum (Splenius) and left occipital lobe are unable to read but can still write (Alexia without agraphia)!

Less commonly, when a part of the brain that normally

causes inhibition is injured, certain behaviors that are normally inhibited can be increased; the classic example is damage to the pre-frontal and orbito-frontal cortex when affected by trauma or a tumor (more commonly Meningioma) can cause antisocial behavior.

Another case that we see more frequently in the neurological practice is the Syndrome of Gertzman, when there is injury of the Angular gyrus of the left parietal lobe, this produces inability to make simple calculations, inability to differentiate left from right, inability to distinguish the digits of the hands, and inability to write.

In all these particular injuries what we see is that the brain definitely controls specific functions of the body and many cognitive abilities of the mind; however, even if you lose your ability to walk, see, coordinate, calculate, or communicate in cases of aphasia, You, your own Self, while experiencing these limitations and disabilities, maintain always your identity and individuality. I have seen patients with large strokes on the left side whose Aphasia was very severe and caused the inability to communicate; however, after months or years, when I talk to the relatives of these patients who have cared for them, they always tell me that they recognize that the person is still their husband or wife, and despite the frustration, they have learned to communicate in other ways such as through visual expression. When they see them, they still know that they are right there and that their individuality remains, their affection and their love remain. A patient who recently suffered Gertzman's syndrome despite her disabilities kept her dreams and concerns and affections intact, and she was still herself.

As mentioned before, patients with disconnection syndromes with separate brains due to surgical procedures of the Corpus Callosum are still able to swim, to ride a bike, and continue to have their own dreams, emotions, affections, and concerns in their lives and maintain their individuality. Even if they face the adversity of neurological disability, they still remain themselves.

Someone could argue that other more diffuse brain injuries such as those of dementia patients or coma patients who are lying in a bed or as patients in Permanent Vegetative State are an indication that damage to the brain results in loss of the person and their individuality as such. I would argue, however, that the demented patient is still that husband or wife or mother of the family who cares for him and that he would react differently to a stranger or to a doctor than he would to his relatives even though his answers are primitive (such as a smile or a rejection) because there is still a connection (although not expressed verbally) with the person who cares for him or her.

Considering patients who are in Persistent Vegetative State, -unable to move or communicate-, one may wonder if it is possible for them to be thinking of something. Well, there are studies that show brain and cortical activity in those patients, and even though we do not have an outside connection, they still have internal activity to which we do not have access. This has been documented with studies of Functional Magnetic Resonance with various stimuli. For example, Dr. Bekinschtein and his Cambridge group were able to find activity in the Premotor cortex in some patients in a vegetative state - unable to move or communicate - when they were instructed to move their left or right hand[xii]; and the same is corroborated in a similar work by Dr. Adrian Owen of Oxford that uses different motor paradigms including telling the patient to imagine that he or she is playing tennis.[xiii]

In the examples above, I have emphasized the permanence of the Individual or the Self as such. This concept is linked to the idea of Consciousness. Consciousness always involves subjectivity (what the philosophers have called Qualia or the subjective experience). This Experience is like a movie in which you are both an actor and the viewer sitting and watching your own movie. In other words, the astonishing reality of consciousness is that we can question and observe ourselves, we are aware of being conscious, and we wonder about this state of being.

Materialism argues that consciousness does not exist separately from the body, not only by the examples we saw of brain damage but also by other circumstances. For example, the fact that the concept of self-being and individuality seems to be lost with the use of hallucinogens could be cited as evidence; however, this only causes a different perceptive experience (in the same way as hunger, fear, and phobia do), and the Self is not lost because the experience is unique to each individual. Despite the same brain connections and same drug that caused the experience, the hallucinations which are induced are different for you than they are for any other person, so your individuality and your self are maintained.

Similarly, the argument arises that during anesthesia, there is no evidence that there is any consciousness or memory of it. However, there have been studies done in which it is documented that some patients dreamed during anesthesia with Propofol and other studies have also been done with different anesthetics such as Sevoflurane[xiv] or Desflurane by a different group.[xv]

Another great topic of discussion has been near-death experiences (NDE). If they are true, they would tell us not only that not all our existence is controlled and limited by our physical body, but also that we also survive after death, so it would speak of the survival of being beyond death. Of course, there is a lot of gossip surrounding this subject, but after a good review of the literature, I will mention some controlled cases. For example, the case of Pamela Reynolds (American singer) who had brain aneurysm surgery. Because it was a brain surgery, there was electroencephalogram monitoring, and it was documented that based on electroencephalogram monitoring at the time of cardiac arrest, there was no brain activity; however, after resuscitation, she explained that she had a near-death experience. The fact that there is no brain activity serves as a counter-argument to those who say that these near-death experiences are only manifestations of Ecstatic seizures. Near-death like experiences in patients with this kind of epilepsy originate in the cortex of the Insula antero-dorsal[xvi]; however, the absence of activity in the electroencephalogram under these circumstances would make this explanation unlikely considering the great detail of the description of the experience. The fact that different cultures report different visions (some cultures see Jesus, others see Buddha - a process called Tibetan Buddhism called "Delock" which means the one who has returned from death- others see Muhammad) does not minimize the more common features of survival to death; immortality and reunion with the deceased relatives no matter the culture.

Some neuropsychologists cite that similar experiences occur with exposure to Ketamine (anesthetic that induces hallucinations), hyperventilation, hypercapnia (increased CO_2 in the blood), or electrical stimulation of the temporal lobe (in patients who are having temporal lobe surgery due to epilepsy). However, the highlight of these experiences compared to those that occur during cardio-respiratory arrest is that in the latter the experiences are very clear and very lively, are implanted in the memory of these patients, and they also turn into transformative experiences of life, which does not happen in former circumstances.

A prospective Dutch study of patients with cardiorespiratory arrest showed that only 18% of these patients had NDE.[xvii] A similar percentage was recently corroborated a larger study: The AWARE Study[xviii] published in Resuscitation Journal in 2014 by Sam Parnia et al. The study was conducted in the UK, Austria, and USA, including centers such as Montefiore Medical Center, Stony Brook, Emory University, Cedar Sinai, and Indiana University. The study involved 2060 episodes of cardiac arrest patients: only 9% of patients experienced NDE and only 2% had experiences outside the body (in which patients had visual memory of what happened in the resuscitation room).

Dr. Jeffery Long, physician specializing in Radiation Oncology (Mary Bird Perkins Cancer Center, Louisiana has also done significant research on near-death experiences.[xix] One of his patients who was a professional singer was congenitally blind (born blind and had severe MVA with multiple injuries. The patient almost died, and during her NDE (near death experience had a vision of herself, her features, her hair, and the ring she was wearing, which was initially very surprising to her since she has never experienced the sense of vision. How can we explain this considering she was congenitally blind if not by something other than the brain function? Also Dr. Long has reported several patients who had cardiac arrest during surgery and under general anesthesia which makes it very difficult to claim that it was the brain causing their visions and NDE which is so clear and very transformative for the patients who had it.

You may argue that this research is just based on what people and patients have reported verbally and that there is no way of proving their account. However, in my opinion, the most revealing is not the fact of the typical near-death experience with the classic description of bright light , a tunnel, reaching a place of infinite peace, and meeting with loved ones; for me what is most convincing of the reality of these experiences is the demonstration that some patients have revealed information that they could not have known unless they were floating and seeing themselves from above in the room. For example, a patient who was blind since childhood was able to describe what she saw in the resuscitation room: who came in, who was wearing what colors, and who was talking. Others have reported written information on a board in the surgery room that was written only after the surgery was already advanced and therefore could not have been seen beforehand by the patient. Dr. Bruce Greyson (University of Virginia) has reported a child who recalled seeing her sister who had died in his vision but was not known to have died by the parents and was in college living in another city. When her parents called her, they learned she had died the day before in a car accident. These latter accounts have no other explanation than the existence of consciousness which is separate from the brain and survives the brain after death.

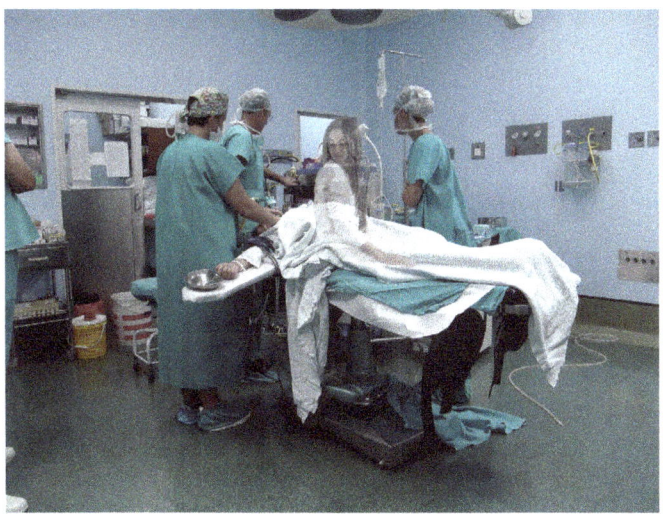

Figure 16. Near Death Experience representation. From Istockphoto.com.

There is another controversial issue which still has many different opinions even today, despite having been a philosophical aspect considered for centuries. Free will, which is a perception that we have the freedom to choose an action or thought, to do something or not, to let an emotion explode or not, to help or not to help, to speak or remain silent, etc. It is a perception, and therefore, can be studied as such.

The pivotal study which brought on all the controversy was done by Libet et al in 1983. He designed an experiment in which he determined the time of the perception of willing a movement and the actual time of the movement. He looked at the electrical brain activity by electroencephalography (EEG) with the Bereitshafts Potential (or Readiness Potential RP, see Fig 17), which corresponds in the electroencephalogram to the locations of a C3, C4, and Pz (central and paracentral parts of the head), negative rising potential obtained by recording Motor and SMA
(Supplementary motor area) electrical activity prior to a voluntary movement.

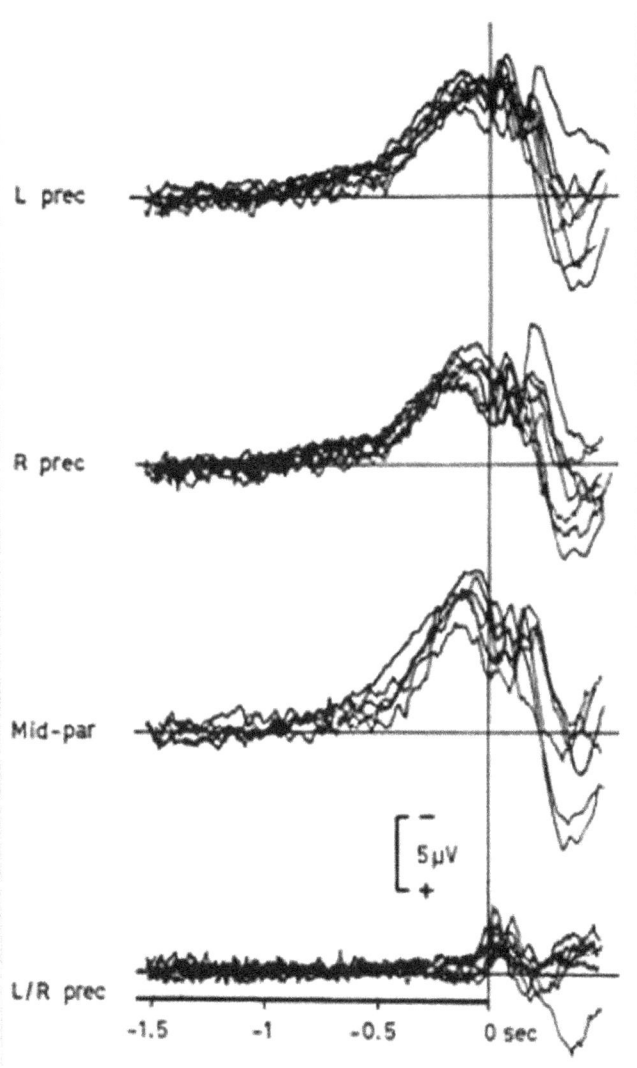

Figure 17. This is the typical, average recording of several traces of the Bereitschafts Potential (translated from German as the Readiness Potential RP) which corresponds to a C3, C4 and Pz electroencephalogram recordings (central and anterior and mid parietal

regions of the head) which occurred prior to voluntary movement (flexion of the Right Index finger) which occurred at the 0 sec time by electromyography recording. This was presented by Hans Helmut Kornhuber and his then doctoral student Lüder Deecke initially in 1965. Note that the activity starts about -1.2 seconds and has a second more defined negative potential (going up) about 0.5 seconds prior to the movement. From Wikimedia Commons. Public Domain.

This potential is 100 times smaller than the regular EEG alpha activity (which is present when you are awake and relaxed with eyes closed) and becomes apparent only by averaging several potentials (about 50 or more) prior to the onset of a movement. This potential has also been recorded by magnetoencephalography by L. Deeke and H. Weinberg in 1982 and has been replicated by recording the medial Frontal cortex during epilepsy surgery on awake patients.[xx]

Libet had the subjects make movements at freely chosen times (flexion of a finger or wrist) and then report the time that they consciously willed the movement by stating the position of a ball on a "clock" that moved around once every 3 seconds. He wanted to measure the subjective feeling of willing the movement and noted that the willing would occur about 300 milliseconds prior to the movement onset.

The RP started about 1 second prior; therefore, the RP started about 700 ms prior to the onset of the willing perception (W) suggesting that the brain initiated the movement prior to the person being aware of willing the movement. Therefore, this finding could be interpreted as the fact that there is no free will and that the brain controls our decisions to move even before we are conscious of them (Fig. 18).

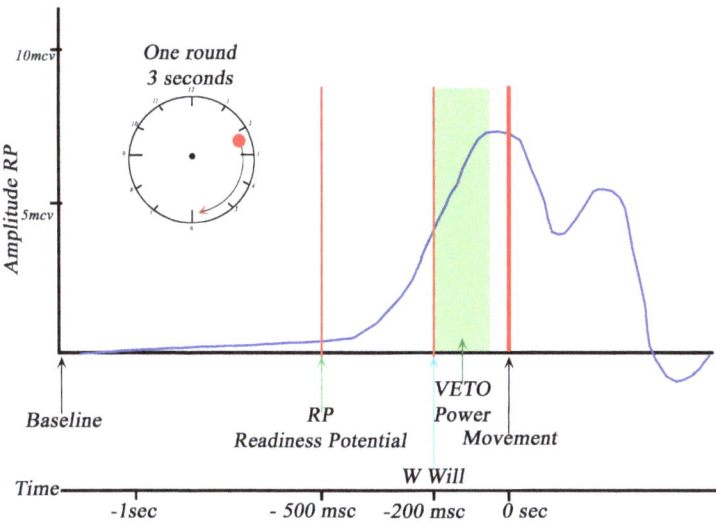

Figure 18. Libet Experiment. Subjects had EEG recordings of the Readiness Potential and were also hooked with surface electrodes recording EMG of the finger and wrist flexor muscles which activated when the actual movement occurred. Subjects were told to move at will, and also, He registered the time when they consciously decided to make the movement by reporting the position of a ball in a "clock" that moved around once every 3 seconds.

Many other experiments after this have refined our knowledge about the subject. By Functional MRI it has been determined that the earliest manifestations of planning a movement could be seen up to 8 seconds earlier.[xxi] The RP has also been seen not only prior to a movement but also prior to a thought.[xxii]

Others such as Matsuhashi and Mark Hallet were able to determine another component called T (thinking about movement) which still preceded the Will but was after the RP, suggesting the following chain of events: unconscious movement initiation, thinking about the movement, willing the movement, and the movement occurring (Fig. 19).

Interestingly, the perception of will could be modified with single pulse TMS (transcranial Magnetic Stimulation) over the SMA-supplementary motor area (not over the motor cortex) in which case the perception of will was reported later. Direct current stimulation in either Angular gyrus would cause will to be reported earlier; and on a different experiment (where a button press caused a generation of a tone which was delayed at different intervals) W could be reported to occur after the actual movement (when the tone was delayed more than 5 ms).

The "Urge to move" could be produced by stimulation at the SMA (direct stimulation of the cortex during epilepsy or Neurosurgical procedures) and the "intention to move" could be obtained by stimulation over the parietal cortex, suggesting that the origin of W (will) is in the parietal cortex and is a perception as noted before.

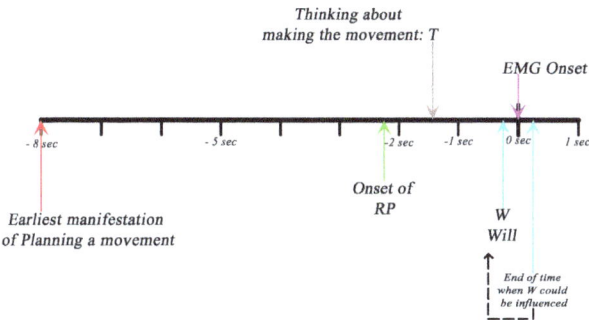

Figure 19. This is the more current line of events from Movement initiation to execution with the associated subjective perceptions (in this case W and T) described by Hallett (Physiology of Free Will, Ann Neurology 2016 Jul: 80(1); 5-12.) As noted from fMRI data, the earliest manifestation of movement could be detected up to 8 seconds prior to the onset of movement. Hallett got onset of the readiness potential earlier than Libet, at about 2200msc prior to movement and they identified another component: T or thinking about moving which was earlier than Will Perception at about 1400msc prior to the movement. The above suggests that movement initiation begins unconsciously, followed by probe awareness and then by full awareness. Noted that Will perception could be influenced and may be reported after the movement already occurred but was referenced back by the subject as occurring prior to the movement. W and T are subjective perceptions and RP and movement are real world events.

The above has to be explained and understood from the real perspective that we live in the past! Our perception of real-world events is delayed from a few to hundreds of milliseconds. After the perception, there is further delay for the processing of that information in the brain. As a mode of example, let's think about the perception of a bright light: light gets to stimulate 6 million cones in the retina which have 3 different varieties sensitive to blue, green, and red; when stimulated by specific wavelengths of light, it reaches via the optic nerve and optic radiations the Occipital lobe which has about 10 billion Neurons and 1 trillion synapses; to get there it takes about 100 ms and then it takes another several milliseconds to reach our awareness and perception. In the same way, a sensation of pinching the thumb may take about 20 ms to reach the Parietal cortex and a sound may reach the auditory Temporal cortex in less than 10 milliseconds. It is this delay which causes us to live in the past so our perceptions are not instantaneous.

Despite the fact that we live in the past, we experience life as present and continuous; the same could be said of the light perception coming from the sun which we receive about 8 minutes and 20 seconds later or light from the Orion Nebula that takes about 1344 light years to reach our retinas and brain.

The same occurs with our perception of will, which is likely determined at the parietal cortex by feedback from the SMA (where a movement is planned) and Motor cortex (where the movement is executed), and as noted earlier, the will perception occurs prior to the movement and at times even after the movement.

It is true that most of our brain activity is unconscious and most movements, events, and actions rarely arrive to our consciousness where there may be a debate or decision making process and then *you may have the power to Veto your decision or movement (this was called by Libet as "**Free won't**"* which also was clearly noted in the Libet experiments and also occurs about 200 ms prior to the movement).

You always have free will because despite the fact that your movements, actions, and decisions are most of all determined by the influences of various parts of the brain including your drives such as hunger, thirst, sex desire, hormones, and neurotransmitters, your geographical or social circumstances, prior actions in the past, memory of prior movements or decisions, feedback from the visual and sensory cortexes as well as thalamic and basal ganglia influences, mirror neurons (Fig. 20) and genetic drives acting on the brain; and in the case of a motor action or movement, while all these influences act upon the motor cortex and the output movement is generated unconsciously at first, you will have the power to veto this unconscious or instinctual drive, to do it or not, to veto it or let it happen.

It is very likely that *this Veto capacity to stop an unconscious or instinctual drive is due to the development of the Frontal lobe* which enables human species to evolve and be in control; it is likely that other mammals and more primary species cannot veto the urge or the drive and that is the reason their actions are instinctual.

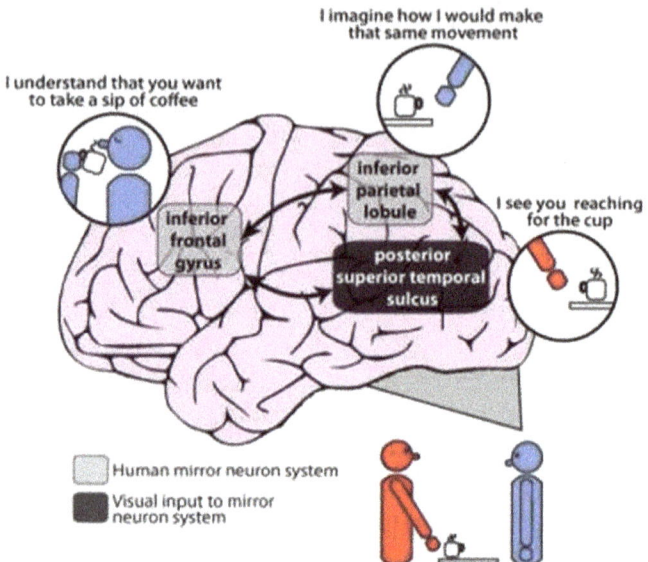

Figure 20. Representation of the Mirror Neurons system. They were initially described by an Italian group (di Pellegrino G., Fadiga L., Fogassi L., Gallese V., Rizzolatti G. "Understanding motor events, a neurophysiological study." Exp. Brain Res. 1992; 91:176–180). They inserted electrodes in the cerebral cortex of Macaque monkeys and found that in an area of the Premotor cortex in the inferior frontal gyrus and also in the Inferior Parietal cortex there is a group of neurons which are active not only when the primate makes an action but also when it is observing a similar action done by another primate or by the human examiner. Later it was found by functional MRI that humans have the same system and the Mirror Neurons were actually found in other areas as in the Primary Motor cortex and the somatosensory cortex. Mirror Neurons are considered to be extremely important for imitation, learning, language, learning the other's intentions and empathy. Dysfunction of the Mirror Neurons have been associated with autism and squizophrenia. From Jan Brascamp. Creative Commons attribution 3.0 license.

The above noted experiments are related to movement and sound which are controlled in a laboratory environment and take only a few milliseconds or seconds. Life is more complicated than this, and many have argued that the laboratory experiments cannot be translated to real life in which you may have a multitude of variants, persons affected besides yourself, and influences of friends and family. As mode of example, if you have to make a decision to take chemotherapy, surgery, and or radiation therapy or not for the most malignant cancer of the brain (Glioblastoma Multiforme), this is not a decision of a few milliseconds and likely would take several days or weeks in the making because it encompasses all your life aspects and existence. A decision like this would make you think about the value of your life and the fact that you are faced directly with death and the end of your existence. You would need to counterbalance the side effects of chemotherapy and intervention which only could add a few months to your existence. Finally, you would need to consider the implications to your spouse or children, the implications to your finances, etc. So, the Readiness Potential (RP) seems truly irrelevant here!

In summary for this topic, we have seen, we live in the past for real events, free will is a perception driven by the parietal cortex, and this perception does not occur before the Readiness Potential and could be manipulated as to occur even after the movement. The brain does most of its functions unconsciously, and behavior could be probabilistic based on current and past life experiences. Yes, this is how your brain works; however, you can veto a movement regardless of whether the RP was already generated unconsciously, and *by having the capacity of vetoing, you disregard that RP and act freely,* contrary to the lack of veto in most other animal species which act by instinct. You decide if you let the physiology act or not, and that sounds to me like you are free to choose and that you are the boss.

What was not addressed is what generates the veto capacity, it may be from the evolutive development of the frontal cortex but also may be related to your consciousness and the subjective experience of your physiology. Maybe your consciousness is inside, above, all around you, and is aware of your brain's physiological mechanisms and connections but at the same time is separated from it and enables you to make decisions and giving you the ability to choose…(Fig. 21.

Figure 21. You have the power to Veto an action and by doing so you have free will. Despite that your genetics, your environment, your culture, the society and parent teachings and prohibitions, the social pressure, your basic physiological needs, hormones, and neurotransmitters push you to act in a specific way, you are able to Veto all that and take another path; you have the power to

choose. From Richard Croft and Peter van der Sluijs. Creative Commons, attribution 2.0 and 4.0 Licenses respectively.

From the point of view of the current Neurophysiology study, there have been several attempts to study consciousness; I will mention a few works which I think have significant value in understanding the physiological processes which are correlated with consciousness. Since the subjective experience or Qualia cannot be understood at this time, most of the studies are done in an effort to study the contrasting findings in the brain from when one is fully conscious to when is not, as it happens in a comatose state or a dreaming state. I will mention some works which are more relevant and valuable in my opinion: The Neuronal Assemblies of British Scientist Susan Greenfield, the Global Neural Workspace initially proposed by Bernard Baars and then elaborated by others including Sid Kouider and the 40 Hz Resonance of Colombian neuroscientist Dr. Rodolfo Llinas.

English scientist Susan Greenfield believes that consciousness arises when there is sufficient stimulus in the brain enough to recruit a larger collection or network of neurons which She calls Neuronal Assemblies. These are groups of 10 to 100 million neurons which are actually the middle way between small group of neurons which represent a focal hardwire connection and large connections among adjacent or more distant areas of the brain. She states that a stimulus for awareness is like a stone in a puddle which starts focal and then spreads like the waves in the water. Any sensory or even cognitive stimulus may be the initial trigger. She states that the personal experiences through life and the different development, even in twins create brain synapses and connections which are unique to the individual and the same stimulus may generate different emotions and responses. The intensity of the stimulus and well as the role of the neurotransmitters which She calls modulators may or may not recruit more and more neurons and create a Neuronal Assembly, the dynamics of the assemblies either shrinking and enlarging determines the level of consciousness. So, in this way a small assembly occurs when you are unconscious, or when you are under the influence of drugs (because you lose your self-awareness and self-control) and a larger assembly would occur when you have depression (intense negative feelings all the time) or during meditation.[xxiii]

Cognitive neuroscientist Sid Kouider has shown that a very brief stimulus (visual or auditory) may activate the primary cortex specialized for the sensory modality without a person being aware of the stimulus. For instance, a brief visual stimulus like the face of a person lasting about 33 ms in duration does activate the Fusiform area (Brodmann 37 in the inferior temporal lobe, which has been clearly determined as facial recognition area), yet the patient is unaware of the stimulus; however, if the visual stimulus is maintained for > 200 ms, there is activation of a larger network which is the Parietal-Frontal cortex which is now known as "Global Neuronal Workspace." This concept was initially offered by psychologist Bernard Baars,[xxiv] which in a way is similar to large Neuronal Assembly (described above by Greenfield) due to large connectivity between the Frontal and Parietal lobes and mediated by long range axons which connect both areas. These studies have been done with EEG and fMRI. Kouider has further elaborated and deepened the concept.[xxv] Similar findings have been noted in babies for their mother's facial recognition (seems an odd question but this is an attempt to answer the question of whether little babies are conscious or not).

Colombian neurophysiologist and professor Rodolfo Llinas has done lots of work from the cell (the neuron) up, sort of a bottom to top approach. Llinas' experiments led to his discovery of the voltage dependent P/Q type Calcium channel.[xxvi] Among his hundreds of publications, in his book titled *I of the Vortex, from Neurons to Self*, he affirms that the ongoing internal activity and the outside activity (sensory input) constitutes the private life and the external world respectively; so the brain is a closed system and a "reality emulator" of the outside world. The brain has ancestral circuits which are genetically predetermined. For instance, the ability to perceive color without learning to do so, these circuits are enriched by our personal experiences as individuals and constitute ourselves, so cognition is an intrinsic "a priori" property of the brain and does not need to be learned, only the content of it has to be learned.

There are some neuronal groups that oscillate in phase producing coherence and simultaneity of activity, and by doing so create resonance which may bring scattered sensory and perceptive elements into an amplified event which Llinas has called the root of cognition. This is similar to the coherence and resonance of cicadas which start as a single chirp and then grow more and more in unison.

Llinas is a pioneer in the neurophysiological properties of the Inferior Olive (which generates the most powerful synapse in the Central Nervous system with the Purkinje cells) and Thalamic neurons both of which have intrinsic oscillatory properties that are supported by the Voltage dependent P/Q type Calcium channel and Potassium conductance at a frequency of 8-12 Hz for the Inferior Olive and in the gamma frequency at 40 Hz for the Thalamic neurons. The mutual Thalamic connections from the cerebral cortex (Fig. 22) which are far larger than they are from the peripheral sensory systems (further supporting the idea of a closed system) constitutes the Thalamocortical system which function is to relay the sensory related properties of the external world to the internally generated motivations, emotions, and memories.

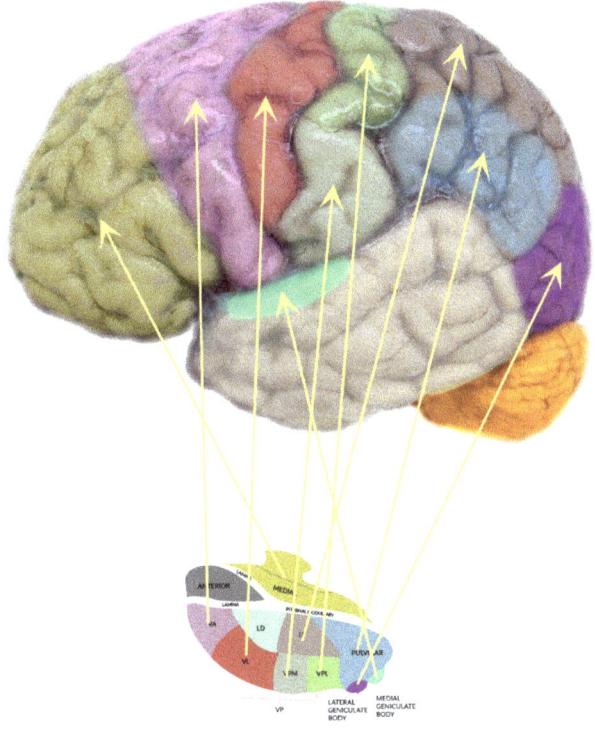

Figure 22. The Thalamo-Cortical system representation. The Thalamus has connections to the entire Cortex. The Thalamus receives sensory information from the external world via the Spinal cord and Cranial nerves, as well as information from the Brainstem, Hypothalamus, Cerebellum, and Basal ganglia (the only stimulus that does not relay in the Thalamus is the Olfactory system). Not pictured here are the projections of the Anterior nucleus to the Cingulate Gyrus. The Thalamus acts as the information central relay and integration system and then transfers to the cerebral cortex which in turn sends feedback to the Thalamus in a two-way reciprocal network (Thalamo-cortical and Cortico-Thalamic). According to Llinas, this allows the external

> and internal information temporal synchronization making the experience be perceived in a continuum allowing consciousness to emerge. The Thalamus has been called the "gate to consciousness."

This temporary coherent event that binds in time the fractured components of the external world and internal reality into a single reality is what he calls consciousness and the Self. He states that the central nervous system and specifically the Thalamus is very similar in fish, turtles, birds, and humans, and due to the 40 Hz resonance of the thalamus (Fig. 23), all noted species have consciousness. He believes single neurons have Qualia or subjective experience. It has been demonstrated, not only in humans but in almost all mammals, that the gamma Thalamocortical oscillation at 40 Hz is noted in the cortex when a predator is stalking a prey and in humans during visual stimulus, auditory stimulus, during increased attention prior to execution of a complex task as well as during REM (rapid eye movement) sleep.

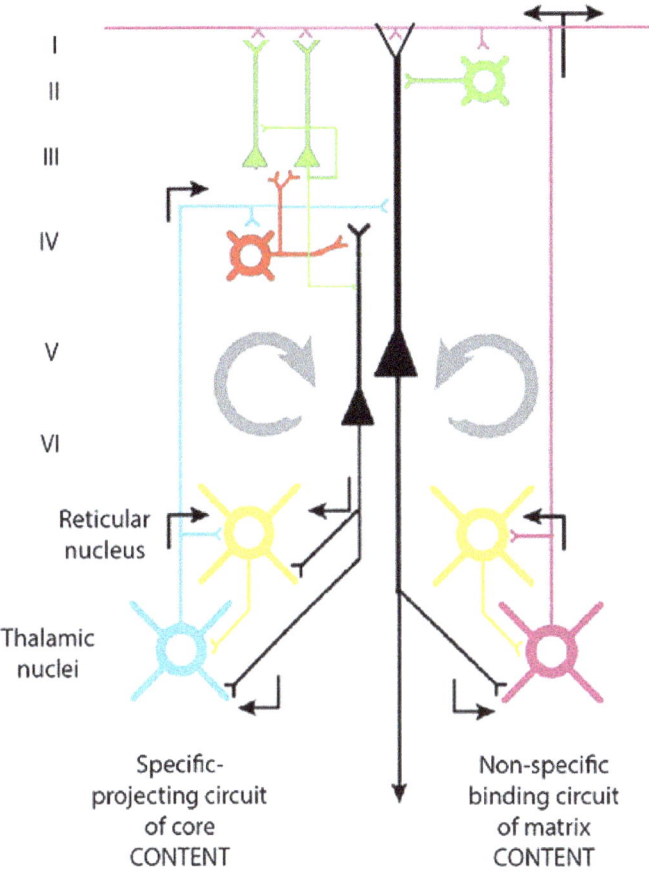

Figure 23. Professor Llinas' proposed diagram of the Cortico-Thalamic-Cortical reverberating circuit which underlines the 40 Hz gamma oscillation noted at the cortex during visual stimulus, auditory stimulus, and situations of elevated attention prior to the execution of complicated tasks. The specific Thalamic Nuclei (blue) carrying specific sensory information (visual, auditory, sensory, gustative, etc.) project to layer IV neurons of the cortex which are GABAergic (use Gamma amino butyric acid as its neurotransmitter) and generate 40 Hz oscillation to the Pyramidal cell (which is excitatory) to generate 40

Hz oscillation as a rebound from abrupt inhibition; these in turn go back and stimulate the reticular nuclei and the other thalamic projecting neurons including the Nonspecific nuclei (purple) which project back to Layer I of the cortex. Llinas hypothesized that the above activity allows a large resonant oscillation between the thalamus and the cortex which provides synchronicity and time-binding of the internal and external world and provides unitary perception and consciousness. With permission from R. Llinas.

Others have tried to understand consciousness from the complementarity of biology with the principles of quantum physics and bringing them to the real macroscopic world. I want to mention the work of Dr. Robert Lanza, who has made known his Biocentrism theory. He states that the Observer is the key for defining our reality; this is true of quantum physics and its main principles which are in a nutshell: matter and wave duality; entanglement (Fig. 24) and uncertainty principles. In all of these real features of the world of elemental particles the Observer defines the reality and is able to determine if light or elemental particles behave as wave or as particles (as in the double slit experiment), two elemental entangled particles get fully determined only when we the observer "looks" at one of the particles and the other gets automatically determined independently of the distance between them and instantaneously (even faster than the speed of light); and one cannot accurately and precisely measure the momentum and position of a sub-atomic particle simultaneously. In all these aspects of quantum physics, the Observer (and by the observer it is meant the electron or photon detector, the polarizing filter or whatever electronic device is used to measure and to observe the quantum properties) causes the probability wave function to collapse and assumes a specific value of reality. The detectors are placed by an intelligent person who of course has consciousness.

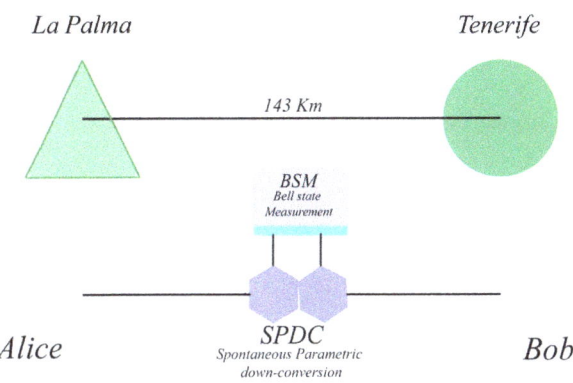

Figure 24. Entanglement. It Is possible to create entangled photons using a crystal of Barium borate. A single beam of light hitting the crystal may split in two beams (this is called down-conversion); since the split beams have a common origin, their properties are entangled. Most commonly with light the property of polarization is used, so depending on the arrangement of the crystal, the outgoing beams can have opposite polarizations (vertical or horizontal); and if you add a second crystal the split beams polarization will be undetermined. Only when **the Observer** measures the polarization of one of the split beams via an electronic detector, the other will assume the opposite polarization and it does it instantaneously (even faster than the speed of light), even if it is located at the other side of the world . Above is a simplified illustration of the group of Viena's work showing entanglement over a distance of 143 Km between the islands of La Palma and Tenerife. The photon entanglement source is generated at the SPDC and then sent to both detectors in the two islands named Alice and Bob and the correlations of the measurements are simultaneously measured by the BSM.
(Teleportation of entanglement over 143 km Thomas Herbst,

Thomas Scheidl, Matthias Fink, Johannes Handsteiner, Bernhard Wittmann, Rupert Ursin, and Anton Zeilinger. PNAS November 17, 2015 112 (46 14202-14205.

Therefore, Dr. Lanza reasons that we—the conscious mind —create and determine space and time (which prior to observation is just a probability wave) and not the other way around. Most classical mainstream neuroscientists believe that the brain creates consciousness (as we noted before it magically "emerges" from brain physiology); Dr. Lanza argues that our brains are actually a construct of our minds in the same way that our heart, legs, kidneys and the external world are a construct of our minds, therefore the mind is not our brain.

His conclusion is that since the Observer is the deterministic entity of the external world at the quantum level and since the observer has consciousness, then, consciousness must be a fundamental part and true nature of the entire universe, and it is different from its physical components. This concept has its own implications, including that there is no death (just an illusion), there is no time, there is no space (since they are modifiable and probabilistic), and we are a single part of the great consciousness of the universe. This is not a fantasy or imagination—it is how the world functions at its very deep and quantum level (these properties of particle-wave duality, entanglement and uncertainty principle has been corroborated in many experiments over the last century). Niels Bohr stated nearly a century ago: "When we measure something, we are forcing an undetermined undefined world to assume an experimental value. We are not measuring the world; we are creating it," and more recently

Stephen Hawkin and Leonard Madinow wrote back in 2010: "There is no way to remove the observer- us- from our perception of the world."

Many people in the physics and scientific community mainstream have criticized Dr. Lanza's Biocentrism theory and its implications. His theory has triggered a revolution in science. Many initial ideas in science which are different from the mainstream are not initially accepted; however, his idea has exactly the same reasoning that Einstein's idea when he postulated his relativity theory. Einstein took really seriously the idea that the speed of light is constant, and therefore, time and space themselves need to be modifiable in the same way that Lanza is taking really seriously the facts already known about quantum physics: the observer is pivotal in determining the reality and is part the equation. The concept of reality that this truth implies needs to be adjusted, even if it takes time for the scientific community to accept it.

All the above arguments lead us to question the main issue: How is it possible that we, who are made of matter, quarks, and electrons (Fig. 25), can reassemble into complex beings in such a way to produce the emergence of a new property of matter called consciousness and the self?

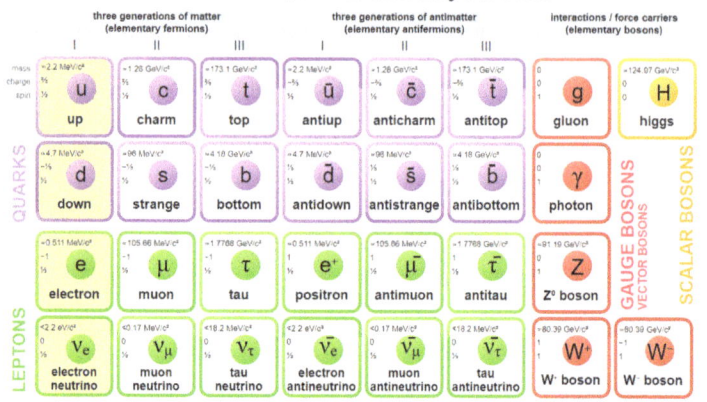

Standard Model of Elementary Particles

Figure 25. This is the standard model of elementary particles at present time. Note that the Higgs Boson was recently discovered in 2012 thanks to the Large Hadron Collider in Geneva. All ordinary matter is made of first generation particles (colored in yellow background on the left Colum) which do not decay (All 3 neutrinos also do not decay; however, they rarely interact with regular matter). Specifically, all matter and all atoms consist of electrons orbiting around atomic nuclei, which are composed by protons, neutrons and are ultimately constituted of up and down quarks. So all that is needed to form regular matter of the universe, including all galaxies, planets, and life, is only the first generation particles.? Why did Nature choose to make another 2 generations of Fermions and even more? Why are there another 3 generations of the corresponding Antimatter (Antifermions)? These are clear questions that remain to be answered. The theoretical Graviton to explain gravitational interaction is not part of the standard Model, although it is suspected to exist as another Boson or force carrying particle. By Johns Hopkins University. 2019. Public domain.

How could something spiritual or non-material emerge or be created out of matter? Is this a new property of an evolved material creature? Is consciousness a primary force of nature and everything including atoms and unicellular and multicellular organisms has consciousness? or is limited to biological beings? Is consciousness an illusion created by the brain, and matter continues to be matter and as such decays when we die and transforms? Or is something instilled or planted by the Creator into the organisms who have reached some degree of evolution? Answering these difficult questions constitutes the hard problem of consciousness as stated by the Australian philosopher David Chalmers (the easy problem is of course demonstrating the physiology of the brain mechanisms, the brain regions involved on an activity of the individual, the pathways and connections of the brain and how they could be either manipulated or intervened to alter the level and degree of consciousness). Our material body, could be thought as the mirror image of our real self or as a shadow of our real identity; it does appear real, and it moves in continuity, it can even see itself in another mirror or reflection, but there is more to it as it is not the real thing.

Our body and all its physiology could also be thought of as a nice modern television set: it provides continuity and appears as reality, and its image gets affected and distorted by changes in all its electronic components, lack of electricity, physical damage, short circuits to one or several of its components; however, the real information (in this example the make of our consciousness and self) is not on the TV which is only a receptor. The real information comes from outside the TV as electromagnetic waves which travel through the air, are invisible to the naked eye, and are indeed difficult to detect.

Chalmers also suggested that consciousness is a fundamental feature of the world like space-time, mass and is integrated with the physical world.[xxvii]

Our daily life is surrounded by non-material properties. we ask ourselves: how do we qualify creativity, empathy, kindness, beauty, love, art, music, imagination, premonition, self-questioning ability, meditation, and many other nonmaterial and spiritual characteristics of human beings which provoke an awe and wonder and speak in favor of a transcendental nature of our species?

I always have wondered about certain characteristics of the self. There is no question that one of the most important aspects of our existence is the affection for family and friends. I would say that emotion towards others: affection, attraction, love, empathy, passion, hate, mark our our existence and determine our pursuance of goals more than any other motivation in life, whether is vocational, professional, or personal growth. I believe that sentiment, emotion and the love between a mother and a son represents a quantitative amount of energy (which is non-material, despite the physiological material manifestations) and is, therefore conserved based in the first law of thermodynamics, the law of Conservation of energy (which states that the total energy of an isolated system is constant; energy can be transformed from one form to another, but can be neither created nor destroyed). This energy, this sentiment could not be lost in the system after the death of the mother or the son. Where would all this love, this energy, have gone? It is certainly not lost at death and transcends it.

There is significant work being done at this time to improve and advance Artificial Intelligence (AI), and even dreams of creating a brain-like computer with consciousness and to that I would say that AI would definitely improve our lives and comfort and would have applications in all aspects of life including medicine. However, is unlikely that a computer will achieve consciousness since consciousness is not just memories and thoughts which represent information, consciousness involves the subjective feeling of existing, how does it feels to be thinking, reasoning, seeing, hearing, touching or smelling, giving or receiving affection, being lonely, desperate or exalted. This is what is not obtainable by computers or artificial intelligence; it may have much better and faster access to information than what we can readily achieve; however, information is different from knowledge which is obtained by experience and feedback and it is only attributed to conscious and sentient beings. A computer may know all about the biology, chemicals, electrical properties, physiology, all genetics and physics of a monkey or a worm; however, it is not the same as knowing what is like to live as a monkey or a worm!

After this brief review, we have seen how a specific damage to an area of the brain, a specific electric stimulation or even a chemical effect can alter your body's function and also your mental capacity. We also have reviewed some of the current mainstream neurophysiological explanations for consciousness either by generating neuronal assemblies, by creating a "global Neural space", or better, by generating oscillations and coherence in the Thalamo-cortical level at the gamma range at 40 Hz all of which may occur with a delay in the perception of time and in the perception of "will" due to the fact that we do live in the past.

We are capable of free will since although we can act at specific time or situation in accordance to our prior physical, genetic, and physiological experiences, we also have the power to veto an action and take a different path.

In addition, We briefly reviewed more recent views which are based on the central role of the Observer in the quantum mechanics principles, and learned that the consciousness of the observer is a primary force that interacts with the elemental particles that constitutes the universe and therefore it is a fundamental part of it.

All of the above explanations of consciousness are mechanistic and give us a plausible explanation of function and physiology, but we get stuck due to the fact that these descriptions are correlations and do not constitute cause-and-effect of consciousness. The mainstream scientific community believes that from these physiological aspects of the brain, consciousness "emerges", " is created", "originates", or "develops", in a magical way. The hard problem of consciousness if approached from the mainstream scientific method will continue to remain as the hard problem due to the fact that this is a category error, which assumes that objective findings of neurophysiology, and the subjective world or Qualia are in the same realm, which they are not, and we are trying to understand a non-material concept with material methods and means.

A logical alternative explanation is that besides our material make up, we do have a "conscious self" that goes beyond our physiology and matter. Although it interacts with it, its nature is not physical and does transcend our material body and our life on earth and continues thereafter.

I understand that at this point there is perhaps no proof or experimental evidence of a non-material consciousness, self or existence, and this has only been glimpsed by those who have personally experienced a close-call or a revelation, which in science is usually dismissed and would amount only to "anecdotal reports". Nevertheless, despite that consciousness has proven so difficult to be studied and defined, it is very close to home in each of us in our daily life. In our inner mind, regardless of our prior beliefs and beyond what is proven and reproducible, you have "talked" to yourself many times, you have observed yourself sitting and thinking about something; or even many times, you would say, " I should have not done this", "I should have done that", "I should have not reacted that way" as if you have your own Self watching over you and making comments about you as a "third" person which you know is your greater SELF (Fig. 26).

Despite the chains of our body, our physiology, and our chemistry and despite the handicaps and frustrating limitations that we suffer, when a part of our body or our brain is affected, there is no doubt in my mind that this is not the entirety of existence and that there is transcendence to our short material life. As stated by the French philosopher and Jesuit priest Pierre de Chardin: "We are not humans beings having a spiritual experience; we are spiritual beings having a human experience."

Figure 26. Consciousness/SELF representation. From Wikimedia Commons Public domain.

Difference between Religion and Spirituality

There is no question that religion has brought many positive contributions to humanity. Many great scientists have been religious people, and many remarkable discoveries in science have been achieved by them. I would mention Robert Boyle who contributed to the gas laws and stated that a deeper understanding of science was a greater glorification of God[xxviii]; James Maxwell who, besides his great contribution to Electro-magnetism and light's electromagnetic nature, also memorized the Bible by age 14; the Augustinian abbot Gregor Mendel gave the foundations of genetics, the catholic priest Georges Lemaitre contributed to the understanding of the expansion of the universe and to Hubble's law, Isaac Newton was also very religious and even Werner Heisenberg, one of the founders of Quantum physics, with his uncertainty principle who was a devout christian and stated in a letter to Einstein: "We can console ourselves that the good Lord know the position of the [subatomic] particles, thus He would let the causality principle continue to have validity."[xxix]

Religion has provided and facilitated many services such as schooling and medical care to the point that presently, many hospitals in the USA and the world carry religious names and their directives and foundations contain, in part, religious principles in their mission description; just to mention a few: Maimonides, St. Francis, St. Mary's, Good Samaritan,

Beth Israel, Interfaith, Jewish Memorial, Methodist, St. Anthony, St. Barnabas, St. Luke's-Roosevelt, Catholic Health, St. Agnes,St. Vicente, St. Cross, Guy's and St. Thomas, Virgen de las Nieves; . Pope Gregory XIII introduced the Gregorian calendar. Many cathedrals and universities in Europe were founded by the church. Religious themes dominated many masterpieces of art for centuries such as the ones by Raphael, Leonardo da Vinci, and Michelangelo. Many great music masterpieces by classical composers were about sacred music, and we still listen to some of them such as Schubert's "Ave Maria." Christianity contributed to the end of many harmful practices such as human sacrifices, infanticide, and polygamy. Epidemiological studies have shown an inverse relationship in religious and spiritual practice versus depression.[xxx]

On the other hand, most of the shift away from religion of the current youth population in the world is not only due to the misbelief that evolution and science are incompatible with the concept of a Creator, but also the fact that in this generation more and more of our concepts are based on what we can touch, see, and experiment with, and scientific facts focus mainly on materialistic aspects of the universe with a frontal rejection and dismissal of anything which is non-physical.

For others, a greater breach has been created by the double morality most religions have shown throughout history. Since religions function under institutions and institutions are led by men, then by looking through history at the leaders and facts of major religions including Judaism,

Islam, and Christianity, one can only be terrified by the horrible acts which have been perpetrated in the name of religion including honor killings, stoning (which still is religious law in some muslim countries), and the Crusades, which went on for centuries (from 11th to 17th centuries) and were considered as "Holy wars" pitting the Catholic Church against Jews and Muslims and vice versa. Since the 13th century, the Catholic Church established the Inquisition to deal with and punish heresy (anyone who did not agree with the church) through jail and torture, which included the rack to stretch and break the body called "strappado" in which the victim was lifted off the ground by his wrists bound behind his back resulting in dislocated shoulders, water torture, and then death by burning at the stake on a public spectacle called "Auto da Fe" (Fig. 27). Historical records state that in 1482, Tomas de Torquemada was appointed as great Inquisitor of Spain, he killed more than two thousand people by the end of the year 1500; also, tens of thousands were killed for being considered as witches (participated in by both Catholic and Protestant churches—Lutheran, Calvinist, Anglican—).

Let me mention a brief historical account of one of the above "Auto da Fe" which occurred in Madrid, Spain in 1682:

> "The officer of the inquisition marched on the 30th of May in Cavalcade to the Palace of the great Square. Of the prisoners 20 men and women with one renegade Mahometan were ordered to be buried. 50 Jews and Jewesses having never before imprisoned and repenting of their crimes were sentenced to long confinement and to wear a yellow cap. Now mass began in the midst of which the priest came from the altar, placed himself

near the scaffold and seated in a chair prepared for that purpose. The chief inquisitor then descended from the amphitheater dressed in his cape and having a miter on his head. After having bowed to the altar attended by some of his officials carrying a cross and the gospels and a book containing the oath by which the King of Spain obliged themselves to protect the catholic faith. The mass begun at twelve and did not end until nine in the evening, after this followed the burning of the 21 men and women whose intrepidity in suffering that horrid death was truly astonishing. The King near situation to the commons rendered the dying groans very audible to him…"

Figure 27. Representation of an Auto-da-Fe by Spanish Painter Pedro Berruguete. Here he depicts Santo Domingo de Guzman presiding over an Auto-da-Fe, circa 1493-1499. The men to be burned at the stake and

the other two waiting their turn were Cathars, a Christian religion in which the idea of two gods or deistic principles, one good and the other evil, was central to Cathar beliefs but was against the monotheistic God of the Catholic Church. Dominicans claim that it is unlikely that Santo Domingo (St. Dominic) ever acted as an inquisitor. From Wikimedia Public Domain.

Religions have also supported slavery and the abuse of women, and, as we all know, have been implicated with sexual abuses which are currently in the news. Many have also considered religion as the institution allowing oppression of the masses by giving people false illusions as put by Karl Marx, the controversial German philosopher from whom socialism later developed, in 1843: "Religious suffering is, at one and the same time, the expression of real suffering and a protest against real suffering. Religion is the sigh of the oppressed creature, the heart of a heartless world, and the soul of soulless conditions. It is the opium of the people." So, it is not unexpected that after all the above noted, the present generation wants no part in religion.

This is in great contrast to the spirituality and morality which was taught by the great few enlightened people throughout history. Behind most religious teachings are the notions to love, love others, and love your neighbor as yourself. This simple principle which is pure and untainted cannot be corrupted by religious institutions or religious leaders. This has been central to many philosophers including Aristoteles, Confucius and to religions like Taoism, Zoroastrianism, Buddhism, Hinduism, and Sikhism. Some have placed as a negative connotation on expressions such as "do not harm" or "do not do unto others what you do not want them to do to you"; however, I think it was better said as accounted in the gospels by Jesus, "Love your neighbor as yourself." He then goes on to answer the question of "who is your neighbor?" with the Good Samaritan parable. By themselves, most religions are based on the belief of a God or supreme being or creator of the universe, yet we can have spirituality in philosophies which do not consider the existence of God. Let us take a brief note of Buddhism in which its central belief is that we suffer because we cling to and crave the impermanent things of our material and mundane world which cannot lead us to attaining real happiness. This attachment produces Karma which ties our existence to Samsara (the cycle of death and rebirth) and in order to free oneself of this suffering and Samsara, a Buddhist follows the "middle way" delineated in the Noble Eightfold Path which could be summarized as cultivating loving kindness and compassion for others and living with the basic principles of morality towards others.

Over the centuries and through history, what religious institutions have done is to distance themselves from the principle of love and replace them with an endless set of rules which over the years has transformed the original principle into a routine of rituals in which people attend mass, synagogue, mesquite, or temple and do a series of prescribed rituals and follow specific rules, but in their daily lives, they have completely lost the essence of the fundamental principle which originated the religion. Many times, the same set of rules becomes a chain so strict that it goes against the accommodating principle of love. Others have become extremely passionate and take their scriptures in a literal way and lose the meaning and essence of the spiritual principle behind it. Many others also under the strong belief that their religion is the only true religion and all the other religions are wrong, find themselves getting into fights, practicing discrimination, and becoming bias toward others and by doing so act completely opposite to the principle of love toward others. I have pointed out the good and the bad aspects of religious institutions with the goal of maintaining the perspective in what really is important to keep and maintain in your own religion: the main principles of love towards your community and your neighbor.

It is likely that this same feeling of criticism of religions is what has set many people apart from religious institutions; yet it is vital and very important not to depart from the great spiritual principles of love for the self and the neighbor, the true reality of the self beyond atoms and quarks, the purpose in life, and the spiritual connection with the Creator. I believe we all have developed our own personal philosophy of life and our own personal religion (independently of the official religion we profess) which have very similar principles for all human species regardless their place on earth or their culture. I believe that throughout history, there have been profoundly spiritual people who have achieved a close connection with the Creator and have shared those thoughts either verbally or in written form for our learning, growth, enjoyment, and reflection. I truly believe we all have in common that seed of goodness and kindness inside which comes from our innate spiritual nature and connection with the Creator; I think this is a true connection among all of us as species regardless of race, religion, color of the skin, sexual preference,or the philosophy we believe in.

I will close with few spiritual notations:

"See yourself in others. Then whom can you hurt? What harm can you do? He who seeks happiness by hurting those who seek happiness will never find happiness (Buddha's teachings from "Dhammapada." Translated by Thomas Byron).

"Science is not only compatible with spirituality; it is a profound source of spirituality. When we recognize our place in an immensity of light-years and in the passage of ages, when we grasp the intricacy, beauty, and subtlety of life, then that soaring feeling, that sense of elation and humility combined, is surely spiritual. So are our emotions in the presence of great art or music or literature, or acts of exemplary selfless courage such as those of Mohandas Gandhi or Martin Luther King, Jr. The notion that science and spirituality are somehow mutually exclusive does a disservice to both" (Carl Sagan, *The Demon-Haunted World: Science as a Candle in the Dark*).

"The greatest disease in the West today is not TB or leprosy; it is being unwanted, unloved, and uncared for. We can cure physical diseases with medicine, but the only cure for loneliness, despair, and hopelessness is love. There are many in the world who are dying for a piece of bread but there are many more dying for a little love. The poverty in the West is a different kind of poverty— it is not only a poverty of loneliness but also of spirituality. There's a hunger for love, as there is a hunger for God" (*A Simple*

Path, Mother Teresa).

Keep your feet on the ground, but let your heart soar as high as it will. Refuse to be average or to surrender to the chill of your spiritual environment. (Arthur Helps).

Just as a candle cannot burn without fire, men cannot live without a spiritual life (The Buddha).

Final Considerations

It is a great feeling when we demonstrate in person a theoretical fact in science. We definitely rely too much on what "others" demonstrate for us, and then we take their demonstrations as fact. However, we can test the main facts with "do it yourself" experiments; for example, we can easily test the effects of gravity by making free fall calculations, we could test the principles of electricity and electromagnetism, we could verify chemical reactions with organic and inorganic compounds, record electrical activity of the brain with EEG, or see the effects of magnetic stimulation of the brain (during my neurology residence, I had the privilege of participating in research on magnetic stimulation of the brain to elucidate basic neurophysiology under Drs. Vahe Amassian and Roger Cracco). It is a little more difficult for us to test the atom and the quantum world, but still we are able to easily test light diffraction and the wave nature of light with the double slit experiment. We may reproduce a Geiger counter or create a cloud chamber to see alpha or beta particles (Helium nucleus and electron), we may take out a telescope to document the fact that the universe is expanding by measuring the red shift in the spectrum of close and far away stars and galaxies with a Star Analyzer filter. All-natural sciences such as physics, geology, astronomy, mathematics, chemistry, and biology have helped over the centuries to give explanations and answers to the ever-present human curiosity and desire for knowledge about the nature and functioning of the world. We should embrace science which gives us a better understanding of the marvelous and fantastic world we live in.

Science, however, should not be the limit in our understanding of the world; our minds should not be restricted only to what the scientific method can reproduce and falsify but also should be open to analyze facts beyond this. Science has become a religion by itself for many people, and there are many materialistic fanatics out there to whom I would say that although they are very brilliant and make us wonder with their astonishing discoveries, they sometimes become nearsighted by excluding everything which is not falsifiable or outside the material world. There are important facts that should bring humility to science and our understanding of the world; just to mention what we know of the ordinary matter in the universe which only amounts to about 5% of the universe, the other 95% is made by dark matter (27% and dark energy (68% of which we know nothing except that it plays a major role in the observation that the universe is expanding faster and faster rather than coming to a holt as expected from gravity. Science continues with a quick exponential progress and will provide closure to many gaps in our knowledge, but is good to keep things in perspective as noted above.

There are many subjects which have no clear classical material explanation; just to mention a few: how can we explain the contents of dreams which many times may be familiar to the dreamer but many times are something unexpected? We know the physiology, the phases, the EEG correlates, and the neurochemistry of sleep, but we do not have an explanation for the content of dreams. How do we explain the connection between a mother and her newborn baby? We all know that even before there is a noise in the monitor, the mother has a presentiment that the baby needs attention. How do we explain the telepathic connection with someone at a distance? A typical case comes to memory: the history of the German psychiatrist Dr. Hans Berger who, by the way, was the first to record EEG (electroencephalogram) activity in humans in 1924. He had an accidental fall from a horse while being a reservist in the army, and he contemplated for a moment certain death. His sister who was many kilometers away had a "feeling" that he was in danger and urged the father to send him a telegraph, and he has described this "telepathy" event as the motivation for going into medicine. Presentiment is something we all may experience at times, but there is no clear explanation for it. Spanish biologist Fernando Alvarez has proven evidence of anticipatory response/presentiment in Bengalese finches.[xxxi]

There are several facts that although not proven by our current scientific method are very ingrained in the history and culture of many traditions. One example is the belief in reincarnation and past lives which is a significant canon in Tibet, China, India, and Indonesia; and in the western world, there are accounts of past life regressions as presented by psychiatrist Brian Weiss in his book *Many Lives, Many Masters,* which has been criticized by the mainstream psychiatric community, but when you read the tape-recorded accounts by the patients on his sessions, it makes you wonder the descriptions and the facts are real. Also, in most classical religions including Christian, Islamic tradition and Buddhism, Jesus, Muhammad, and Buddha perform "miracles," and classical examples are levitation, multiplication of food, and healing. All these have historical verbal and written accounts. These same miracles are not considered miracles by the Yogis but natural power achievements (Vibhuti Pada – Yoga Sutras of Patanjali). All of the above is seen with skepticism by the scientific mainstream, but its historical pervasiveness in all cultures makes you consider that they are real.

In this short account, I wanted to refresh your awareness on the fact that we are surrounded by transcendental experiences in our daily lives which are difficult to account only by the molecular and physiological patterns of our material brain. Evolution as noted does not exclude the presence of the Creator who launched and initiated the evolutionary process, which over a very long period of time originated the presence of more evolved beings including mammals who are endowed with conscious experience, self-awareness, and subjective experience most exemplified in the greatest achievement of evolution: ourselves, conscious beings who have the ability to take full determination of our decisions and have the choice of independent actions *regardless if we called Free Will or Free Wont*.

The more we go in the microcosm at the quantum level with its weird properties, we continue to learn that we are part of reality and that we influence reality when we interact with nature, so our consciousness is an interactive part of our material world, therefore, consciousness may be considered a fundamental part of reality, not in a dualistic manner but as an integral part of our existence. The self and our consciousness provide us a deeper meaning to our existence, give us purpose and sense beyond our bodies.

Many have stated that the universe is fine-tuned for the existence of life on Earth, and I would say that it is not only fine-tuned for the existence of humanity but also for the existence of many animals such as the dolphins or whales which presently appear at its maximum as species, and why not to say that the universe is fine-tuned also for the existence of many different forms of life on other planets and galaxies? Some beings may have less and some may have more evolved consciousness and capabilities than we have, and maybe all evolution is marching forward to join the Creator, the source.

I believe there are many aspects of spiritual existence which cannot be proven with the scientific method at the present time. They only become real to the individual who has experienced something out of the norm such as a near death experience; this is just anecdotal reports for the scientist, while for others it remains a question of faith, faith in the spiritual nature and transcendence of our existence based on the glimpse of the non-material aspects of our daily lives (love, beauty, kindness, wonder, empathy, music, self-questioning ability). In the same way, we have to use faith also to assume scientific beliefs which are not testable by the scientific method either (e.g. we cannot go back, witness and test the miraculous formation of the first replicable cell 3.500.000.000 years ago, nor can we go back to the Big Bang period or even before that event and make observations). Both of these spiritual and scientific beliefs are not falsifiable, we just have to believe, or not to believe in them as a matter of faith.

As death has "visited" all our homes (as noted in the Buddha history of Kisa Gotami, see **Appendix**), most of us have had one or more close relatives who have died. Some of us had the experience of being at the funeral close to the corpse. The perception and the feeling is that the person is no longer there, "He/She is not there," in other terms, the material body itself appears as an empty shell and is not what we are; the spark of consciousness that made that person is not there anymore, is somewhere else. So as in the case of Kisa Gotami's history there is no permanence for the body, but as the Dalai Lama stated from the Buddhist perspective, "Mind enjoys a status separate from the material world; the mental realm cannot be reduced to the world of matter, though it may depend upon that world to function."[xxxii]

There is only one moment in which we will know the truth with absolute certainty: at the time of our death when the fulfillment of our persistent drive for knowledge and answers and our appetite for the infinity will be fully filled. There will be no more doubts, no arguments, and no uncertainty, just pure knowledge of the truth. Maybe it will turn out to be a sudden cease of existence, maybe a metamorphosis like in the case of a butterfly, a new existence as when life in the utero ends and life as newborn begins, and most certainly a pleasant surprise for the proclaimed materialistic person. I do feel though that the existence of our affection and love for others is definitely an immaterial energy which by our own laws of thermodynamics will not cease to exist!

In the meantime, let us cultivate our full existence, let us embrace science but include spirituality; let us keep attached to the fantastic world of science but maintaining an open mind to what lies beyond with critical perspective but without limiting and blinding ourselves to transcendence.

Appendix

In the time of *the Buddha*, a *woman* named Kisagotami suffered the *death* of her only *child*. Unable to accept it, she ran from person to person, seeking a *medicine* to restore her child to life. The Buddha was said to have such a medicine.

Kisagotami went to the Buddha, paid homage, and asked, "Can you make a medicine that will restore my child?"

"I know of such a medicine," the Buddha replied. "But in order to make it, I must have certain ingredients."

Relieved, the woman asked, "What ingredients do you require?"

"Bring me a handful of *mustard seed*," said the Buddha. The woman promised to procure it for him, but as she was leaving, he added, "I require the mustard seed be taken from a household where no child, spouse, parent, or servant has died."

The woman agreed and began going from house to house in search of the mustard seed. At each house the people agreed to give her the seed, but when she asked them if anyone had died in that household, she could find no home where death had not visited—in one house a daughter, in another a servant, in others a husband or a parent had died. Kisagotami was not able to find a home free from the suffering of death. Seeing she was not alone in her grief, the mother let go of her child's lifeless body and returned to the Buddha, who said with great compassion, "You thought that you alone had lost a son, the law of death is that among all living creatures there is no permanence."[xxxiii]

Endnotes

[i] Brodman 41, 42.

[ii] *What Science Reveals About How Meditation Changes Your Mind, Brain, and Body* by Daniel Goleman and Richard Davidson

[iii] "Temporal Dynamics of the Default Mode Network Characterize Meditation-Induced Alterations in Consciousness." Panda R, Bharath RD, Upadhyay N, Mangalore S, Chennu S, Rao SL. Front Hum Neurosci. 2016; 10: 372.

[iv] Saey TH. 17 September 2018. "A recount of human genes ups the number to at least 46,831". Science News.

[v] Drake JW, Charlesworth B, Charlesworth D, Crow JF, April 1998. "Rates of spontaneous mutation". Genetics. 148, 4: 1667–86.

[vi] Wang, Xiaoxia; Grus, Wendy E.; Zhang, Jianzhi (2006). "Gene Losses during Human Origins". PLoS Biology. 4(3): e52.

[vii] Stedman HH, Kozyak BW, Nelson A, Thesier DM, Su LT, Low DW, Bridges CR, Shrager JB, Minugh-Purvis N, Mitchell MA, March 2004. "Myosin gene mutation correlates with anatomical changes in the human lineage". Nature. 428 (6981): 415–8.

[viii] Gen Bank. "Severe acute respiratory syndrome coronavirus 2 isolate Wuhan-Hu-1, complete genome," NCBI Reference Sequence: NC_045512.2

[ix] Racaniello VR, Baltimore D, November 1981. "Cloned poliovirus complementary DNA is infectious in mammalian cells". Science. 214 (4523): 916–9.

[x] "Creation of a Bacterial Cell Controlled by a Chemically

Synthesized Genome". Gibson, D.; Glass et all Science. 329 (5987): 52–56.

[xi] "Synthetic microbe has fewest genes, but many mysteries," Robert F. Service, Science 25 Mar 2016: Vol. 351, Issue 6280, pp. 1380-1381.

[xii] Front Hum Neurosci. 2011; 5: 5.

[xiii] "Detecting awareness in the vegetative state." Science 08 Sep 2006: Vol. 313, Issue 5792, pp. 1402.

[xiv] "Dreaming under anesthesia: is it a real possiblity? Investigation of the effect of preoperative imagination on the quality of postoperative dream recalls," Judit Gyulaházi et all. BMC Anesthesiol. 2016; 16: 53.

[xv] "Dreaming and Electroencephalographic Changes during Anesthesia Maintained with Propofol or Desflurane," Kate Leslie et all. Anesthesiology 2009; 111:547–55.

[xvi] "Ecstatic Epileptics: A Glimpse into the Multiple Seizure Roles of the Insula." Markus Gschwind and Fabienne Picard. Front Behav Neurosci.2016; 10: 21.

[xvii] "Getting comfortable With Near-Death Experiences: Dutch Prospective Research on Near-Death Experiences During Cardiac Arrest." Pim van Lommel, MD, Mo Med. 2014 Mar-Apr; 111(2): 126–131.

[xviii] AWARE—AWAreness during REsuscitation—A study

[xix] "Near-Death Experiences Evidence for Their Reality." Jeffrey Long, MD, Mo Med. 2014 Sep-Oct; 111(5): 372–380.

[xx] "Internally Generated Preactivation of Single Neurons in Human Medial Frontal Cortex Predicts Volition," Itzhak Fried, Roy Mukamel and Gabriel Kreiman. Neuron 69, 548–562, February 10, 2011.

[xxi] Soon CS, Brass M, Heinze HJ, Haynes JD. "Unconscious determinants of free decisions in the human

brain." Nat Neurosci. 2008;11(5):543–545.

[xxii] Alexander. P. et al: "RP driven by non-motoric process, consciousness, and cognition," 2016; 39: 38-47.

[xxiii] "The Features and Functions of Neuronal Assemblies: Possible Dependency on Mechanisms beyond Synaptic Transmission," Antoine-Scott Badin, Francesco Fermani, and Susan A. Greenfield· Front Neural Circuits. 2016; 10: 114.

[xxiv] Baars, B. J. 1989. "A cognitive theory of consciousness." Cambridge, UK: Cambridge University Press.

[xxv] "What Is Consciousness, and Could Machines Have It?," Stanislas Dehaene, Hakwan Lau, Sid Kouider· Science, 2017 Oct 27;358(6362):486-492.

[xxvi] Llinás R, Sugimori M , August 1980. "Electrophysiological properties of in vitro Purkinje cell somata in mammalian cerebellar slices". J. Physiol. 305: 171–95.

[xxvii] "Consciousness and its place in Nature," Published in S. Stich & T. Warfield, eds, Blackwell Guide to Philosophy of Mind, Blackwell, 2003.

[xxviii] *Disquisition about the Final Causes of Natural Things*, 1688.

[xxix] Gerald Holton, 2005, *Victory and Vexation in Science: Einstein, Bohr, Heisenberg and Others*, pp.32; Harvard University Press, London.

[xxx] "Religious and Spiritual Factors in Depression: Review and Integration of the Research," Raphael Bonelli, Rachel E. Dew, Harold G. Koenig, David H. Rosmarin, and Sasan Vasegh· Depress Res Treat. 2012; 2012.

[xxxi] Journal of Scientific Exploration 24(4):599-610 · December 2010.

[xxxii] Excerpt from "The Universe in a Single Atom", by His Holiness The Dalai Lama.

[xxxiii] Excerpt from "The Art of Happiness", a Handbook for Living, by His Holiness The Dalai Lama and Howard C. Cutler.

www.ingramcontent.com/pod-product-compliance
Lightning Source LLC
Chambersburg PA
CBHW042117100526
44587CB00025B/4086